AF893001

Ferrocement Boat Construction

FERROCEMENT BOAT CONSTRUCTION

Chris Cairncross

International Marine
Publishing Company
Camden, Maine

International Marine Publishing Company
Camden, Maine
First published in Great Britain 1972 by Adlard Coles Limited
3 Upper James Street, London W1R 4BP

Copyright © C Cairncross 1972

All rights reserved. No part of this publication may be reproduced, stored in a retrieval system, or transmitted, in any form or by any means, electronic, mechanical, photocopying, recording or otherwise, without the prior permission of the publisher

ISBN 0–87742–024–6
Library of Congress Catalog Card No. 72–76553
Filmset by Keyspools Ltd, Golborne, Lancashire
Printed in Great Britain by C. Tinling & Co. Ltd., London and Prescot

CONTENTS

LIST OF ILLUSTRATIONS

Photographs

Drawings

ACKNOWLEDGEMENTS

In the collecting of material and the writing of this book I have called on the assistance and advice of many people and organisations throughout the world. But for their help and the encouragement I received from many unsolicited corners I would never have been able to complete this manuscript let alone offer it for publication. To all I would offer my sincere thanks.

In particular I would like to thank Austin L. Edge of the Cement Marketing Co. Ltd, on whom I leant heavily for advice and information. Austin has in one way and another probably been involved with the construction of more ferrocement boats in this country than any other individual. His practical experience and willingness to reveal all he had learned was of great help. To Maurice Griffiths GM, who generously agreed to contribute his thoughts on ferrocement design, I also offer my thanks. To Martin E. Iorns of Fibersteel, West Sacramento, who gave me a breakdown of the situation in the US and the benefit of Fibersteel's experience in this area of marine construction. To John Hodsdon of Ferro Cement Marine Services at Burnham-on-Crouch, who generously supplied me with photographs illustrating different phases of construction for inclusion in the book. To boatbuilder Richard Hartley of New Zealand, who readily answered questions concerning his methods of construction and for clarification of some of the contradictions which litter the published writing concerning this method of boatbuilding.

ferrocement 1

INTRODUCTION

Probably the most logical way to start off this exercise is to ask, simply: what is ferrocement? Essentially one can describe it as steel-reinforced concrete. The term refers basically to a thin shelled structure consisting of a close-knit matrix of small diameter high-tensile steel rods sandwiched between layers of wire mesh, and imbedded in a rich cement mortar which just covers the reinforcement.

It is a 'super' reinforced concrete. In structure ferrocement differs from conventional reinforced concrete in that the steel reinforcement predominates, rather than the concrete, in which the steel is merely a strengthening adjunct. By altering the mortar/steel ratio, to produce a ferrocement structure, we find that these two components actually surrender their separate identities, uniting to form a homogeneous material with an entirely new identity. This new material reacts as a single unit with properties superior to those exhibited separately by the steel and mortar.

Ferrocement is not a new development, nor is its application to boatbuilding a new concept. To review its origins as a boat-building material we would have to go back more than 120 years. It did, in fact, herald the advent of steel-reinforced concrete as a structural material. And records indicate that its first application was for the construction of a boat. As a contemporary construction technique it dates back to the 1940s, when it was 're-introduced' to the world in a more sophisticated form by the Italian engineer Prof. Pier Luigi Nervi. It was Prof. Nervi who coined the term ferrocement.

The principal techniques employed by today's ferrocement boatbuilders have been established for a long time; they differ but little from the methods developed more than 25 years ago by Prof. Nervi. Yet it is only in the last 10 years or so that the main activity in ferrocement construction has been concentrated. There are a number of possible reasons for this late development of the material.

Until a few years ago, so it seems, most people when first presented with the concept of building a boat in ferrocement treated the idea with a certain amount of scepticism and disbelief. No doubt they thought of the material in terms of the conventional type of reinforced concrete. Perhaps the word ferrocement evoked in them images of those concrete land-based structures they were familiar with. Massive clumps of masonry, heavy and brittle. Certainly not the sort of properties one would have thought ideal in a material to be used for boatbuilding.

Or, on the other hand, perhaps people found it difficult to accept that a thin shell structure of mortar and steel would be sufficiently impermeable, strong and impact resistant, or with the right degree of flexibility for this application. Whatever the reason might have been you may be sure that it certainly created a great deal of reticence towards accepting ferrocement as a suitable construction technique for boats. And there is every good reason to believe that it was this reticence which has stunted any wider use of the material in naval architecture until the present time.

However, there is increasing evidence to indicate that the climate of opinion towards ferrocement is changing rapidly. This is clearly reflected in the increased application of ferrocement to boatbuilding which has taken place during the last decade. Construction projects have sky-rocketed from a few isolated experiments carried out mainly by backyard amateurs into a considerable industry. And early forecasts are that ferrocement techniques could in the not too distant future be playing a dominant role in the boatbuilding industry, particularly in the construction of larger sized craft. There are many naval architects who accept that the application of this material has a tremendous growth potential.

As yet the main activity is concentrated among the home building fraternity, for ferrocement has particular advantages for the amateur working on a low budget. The number turning to this medium in preference to the more conventional materials such as wood, GRP or steel is visibly swelling.

This concrete boatbuilding boom can also be seen to be rapidly spreading to the commercial sector. Already there are a number of factories, in various parts of the world, specialising solely in ferrocement techniques. And there are regular reports of other commercial ventures in the process of being formed. Hundreds of professionally built concrete craft, ranging from fishing trawlers to workboats, luxury yachts and motor cruisers, have come off the production lines and are in service.

We could well describe the 1960s as the decade during which ferrocement finally achieved recognition from the boatbuilding establishment as a suitable construction technique. As a measure of this acceptance Lloyd's Register of Shipping are now prepared to issue a 100A1 classification for ferro boats which have been properly constructed under supervision in the right conditions. A couple of British built craft have already been thus classified.

Other professional bodies which are prepared to approve ferrocement are the FAO (Food and Agricultural Organisation of the United Nations) and the British White Fish Authority. The latter is prepared to consider applications for grants for the building or purchase of ferrocement craft, while officials of the FAO have expressed 'tremendous interest' in ferrocement, and see 'fantastic economic advantages in using this material'.

Naval architects and a master builder of the FAO, in cooperation with the ferrocement pioneer Prof. Nervi, recently completed the building of two ferrocement fishing vessels in southern Italy. The boats are for a FAO/UN Development Fund project on Lake Nasser in the United Arab Republic. At the same time, their construction provided demonstration material for the FAO, which had the whole project recorded on film from start to finish. The idea is to use the film for demonstration purposes in developing countries where use could be made of ferrocement techniques.

This growing activity and interest in ferrocement is also affecting the attitude of many established naval architects who have, up until now at any rate, tended to ignore ferrocement. Most of them regarded the technique as 'unprofessional', more suited to the inexperienced builder on a low budget, knocking together a boat in his back yard. Their view was that it was a 'commercially unattractive' material for which to spend time drawing up a design. The result: until recently a dearth of designs specifically drawn up for ferrocement methods. But there are now signs of activity coming from the drawing offices of these architects, and more and more designs are becoming available.

Three countries, incidently all members of the Commonwealth, are much in the forefront of the current spate of ferrocement development. These are New Zealand, Britain and Canada. Strangely enough, in France and Italy, where the material was first developed and later re-introduced, there has been little evidence to indicate any sort of gravitation toward this construction technique from the boatbuilding fraternity; although a number of ferrocement craft have been built in these two countries.

In the United States as well, professional boatbuilders have been slow to jump on the ferro bandwagon. But there are sure signs now that the technique is rapidly taking hold there as well. One or two companies are carrying out intensive development programmes. Perhaps the better known of these is a company called Fibersteel Corp., of Sacramento, California, which is carrying out research into improved building techniques for ferrocement hulls for use in commercial boatbuilding. At the present time commercial expansion in America of this form of boat construction has been going through a difficult patch. High labour costs have been at the centre of the trouble. According to Martin Iorns, President of Fibersteel, 'many would-be builders, entranced by its [ferrocement] low material cost and manifest superiority over fibreglass, have rushed into the field without adequate working capital'. A North American Ferro-cement Marine Association was recently formed to promote the advancement and dissemination of ferrocement knowledge and practices.

In terms of output, and perhaps technology as well, New Zealand holds a slight lead over Britain, which is second in the field. Production-wise, Canada is still way down the line. New Zealand has often been referred to as the home of ferrocement boatbuilding, and not without reason. There is a great deal of enthusiasm there for this form of construction, and a great many boats have been built, or are presently being built. A recently returned visitor to New Zealand remarked to me that 'the home without a concrete boat being built in the back yard is the exception rather than the rule in New Zealand'. This is an exaggeration, no doubt, but it does reflect the attitude of the Kiwis to this medium. Local media, boating publications in particular, and those covering the construction industry in general, have devoted much space to promoting the growing ferrocement industry and acting as a clearing house for any new information or discussion on the technique. A New Zealand Ferrocement Marine Association has been formed, which is doing a valuable job making available information about every aspect of ferrocement marine construction.

This form of boat construction has become one of the most important divisions of the New Zealand boatbuilding industry, with a number of commercially viable yards specialising in ferrocement. Two of the better known companies are Hartley's Boat Plans Ltd and Ferro-cement Ltd, both of Auckland. A great deal of money is being injected into the industry. In an article in *Sea Spray* magazine (1970) one ferrocement expert, Gainor Jackson Jnr, conservatively estimated that well over one million dollars (NZ) had been invested in concrete boats there over the past two years.

1 A 30 ft Seacrete work hull with Lister engine installed. One of three sent to Kenya for local completion.

Britain must undoubtedly have the highest commercial output of concrete boats after New Zealand. And production here is increasing so much that she must surely be closing the gap. Some local ferrocement builders feel that Britain has already taken over the lead in production, and is likely to pass the New Zealanders in terms of technological know-how as well. I feel that this latter view is overly optimistic, however.

Credit for most of the ferrocement craft built in this country must surely go to one firm, Windboats Ltd of Wroxham on the Norfolk Broads. In 1959, under direction of the Hagenbach family, they first began investigating the possibilities of ferrocement. Two years later they built their first boat, and soon afterwards they patented their method under the name Seacrete. In general their techniques differ but little from those adopted by the majority of ferrocement builders. But they have built up a considerable reputation for reliability, and have certainly put the technique on the map in this country. To date they have constructed well over 200 boats, and there are more than 20 of their licencees operating in various parts of the world.

Windboats are not the only company working with ferrocement in this country, but they are, as yet, the largest. In the past year or two a couple of other companies have set up business, but only in a very limited way. There are indications, though, that this form of construction is going to develop considerably here. One well known naval architect mentions a large company with considerable capital behind it which is contemplating branching out into this field. On the amateur side, activity is escalating, and I have knowledge of at least 20 ferrocement boats which have been or are still in the process of being constructed.

Canada has shown evidence too of being involved in this latest ferrocement revival. In terms of output they lag behind New Zealand and Britain. Samson Marine Design Enterprises Ltd of Vancouver, developed by John Samson, is the leading name in ferrocement there. This company has built a wide range of concrete boats and has produced a series of designs. They have expanded their activities to dealerships or agents, who represent them in many part of the world, including America and Ireland. In Vancouver they hold regular courses for newly appointed dealers on their methods of construction.

Other countries in which ferrocement methods are being made use of in commercial ventures are South Africa and Hong Kong. These are either through dealerships or independent builders. A couple of projects are also beginning to gell in some of the newly independent African states as well. In Hong Kong there has been official encouragement for the building of ferrocement craft. Local fishermen are often granted loans, sometimes up to 80 per cent, to assist them in having a concrete vessel built. At the present time they are having their craft built by Ferro-cement Marine Construction Ltd, an associate company of New Zealand Ferrocement Ltd. The most recent vessel to be completed there is a 54 ft fishing boat, built to one of the standard designs of the Hong Kong Agricultural and Fisheries Department. The design was merely adapted for ferrocement construction.

Communist China has been using ferrocement methods for almost 10 years. It is interesting to note that the Chinese are thought to have produced, to date, the largest tonnage of ferrocement craft. These, as far as is known, have been limited to three sizes of sampan, ranging from 3 to 5 tons displacement. The craft operate on the complex system of waterways feeding China's Grand Canal. The factory producing them is situated west of Shanghai and began operations in 1963.

Interest in ferrocement has also been stimulated in Japan, Germany and Ceylon. I understand that Japanese and German interests are to invest in a ferrocement boatbuilding project in Ceylon. A newly formed company there, Manicrete (Ceylon), subsidiary of the influential Collette's Group, is soon to begin manufacturing concrete boats in Ceylon under licence to Windboats of England. This particular venture, it has been reported, has aroused the interest of industrialists from Japan, South Korea and the Philippines, who have expressed a desire to develop similar ventures in their own countries.

It certainly looks as if a successful future is assured for the ferrocement technique in the marine construction industry. It is interesting to speculate what may have influenced this comparatively recent revival of the technique, which has only become manifest in the last decade. Principle techniques have long been established, and there have not been any significant technological advances which might have stimulated growth.

One theory could be that the high and escalating costs of boatbuilding with conventional materials may have caused designers, marine engineers and prospective boat owners to look for alternative methods of construction. And ferrocement has long had a reputation as a low cost technique. At least, the materials are relatively inexpensive. For example, I was recently involved with the construction of a 65 ft ferrocement ketch which was launched in late 1970. A rough estimate of material costs for hull and deck came to £1,200. In this instance the deck was marine ply supported on mahogany laminate beams. If a ferrocement deck had been adopted instead costs could have been lowered appreciably. A similar vessel to the one above, but some 2 ft shorter, was estimated to have cost

2 Mainland China has been using ferrocement methods commercially since 1963 to construct sampans. Factories building them are situated at Wusih west of Shanghai. Six sizes of sampan are built for use on the network of waterways feeding into China's Grand Canal. Construction methods adopt a male mould onto which the reinforcement is laid, as can be seen in the picture. Photo: Concrete Products, courtesy of *Yachting Monthly*.

just on £1,000 for hull materials. Another example is a 40 ft cruiser/racer hull completed last year by Manchester Ship Canal pilot John Law: he estimated that cost of the bare hull, without deck, came to about £250 for materials.

Of course the price of component materials can vary considerably with locality. In some countries, say Africa, where steel may not be as readily available as here, ferrocement methods might not be all that much cheaper. Samson Marine Design Enterprises of Canada provide further evidence of the low construction costs of ferrocement. Accompanying a brief summary of their designs, which include ocean-going yachts, houseboats and motor cruisers, is an estimate of what hull and deck material costs would be there. Here I should add that the estimates were made some time ago, and prices have risen since. They quote 900 dollars (Canadian) for hull and deck materials for a 34 ft centreboard sloop which they call C-Spray; 1,000 dollars for a 36 ft ketch, the C-Shell. Builders in New Zealand and Australia provide figures which fall into line with those quoted above. Hartley's quote 750 dollars (NZ) for hull and deck materials for a 45 ft ferro concrete motor sailer; and about 500 dollars (NZ) for a 37-footer. Costs for the same craft, or their hull materials, in Australia are slightly higher. I must emphasise that these estimates refer only to the costs of hull and deck materials, and that hull costs, even when the boat is professionally constructed, only contribute about one third of the total cost of the finished vessel.

From a commercial standpoint it should be mentioned, as a warning to those who may be contemplating starting up professionally, that while on paper and in the amateur's view ferrocement provides a relatively cheap construction technique, it can be otherwise. This has been well illustrated in the US, where commercial concerns have gone bust with too little capital to back them up. Ferrocement has low material costs, but the techniques as generally used today are labour intensive. And labour, particularly in places such as the US, Canada and Australia, is a costly commodity. It is not particularly cheap here in Britain either. The American company Fibersteel claims to have come up with a partial solution to this problem, using cavity moulds, which may eventually lead to more commercial boatbuilding.

What is undoubtedly escalating the swing towards ferrocement is some of the other advantages it has shown, besides cost, over wood, steel or GRP. Provided the correct construction practices are adhered to, ferrocement hulls have proved to be extremely durable with little demand for continuing maintenance. A good illustration of the lasting properties of this medium,

3 *Featherstone,* a 28 ft ferrocement yawl, photographed at her berth in Fort Lauderdale, Florida in 1968. The yacht was built in 1960 and has been in continuous use ever since with her hull showing no sign of deterioration in that time. The yacht was constructed by architects Larry Browning and George Polk, marine architect Dave Peebles and engineer Walter C. Harry during spare weekends over a period of two years. The cost was roughly 3,600 dollars. *Featherstone* was formed of $\frac{3}{16}$ in. steel rods sandwiched between four layers of fine wire mesh coated with 4,000 p.s.i. cement mortar. With a waterline length of 21 ft and a beam of 7 ft 2 in. the yacht weighs around 5,500 lbs. Photo: M. E. Iorns, Fibersteel Corp.

which will be discussed in the following chapter, is a Dutch-built scow, which first went into the water in 1887. It was still in use three years ago, on a pond in Amsterdam Zoo, until being placed in a museum.

Ferro hulls do not rot, nor do such marine animals as borers and teredo worm have any effect on the mortar. At one stage this was thought not to be the case, and that there was an animal in existence which might have a deleterious effect on a ferro hull. This rumour arose after the discovery of a type of rock clam or borer, known as *Penitella penita,* which was discovered in the mortar jackets of wooden piles in Los Angeles harbour. However, there is little possibility of this animal causing damage. It requires a soft material to bore into, which ferrocement most definitely is not. It would also require a roughish surface for settlement, which should not be present on a hull. And in any case the surface coating of paint or epoxy would be an added deterrant.

Ferrocement boats have shown that they can be far stronger than similar sized production boats built in wood or glass fibre. A particular characteristic of this material, unlike others, is that it gains strength with age, rather than loses it.

There have been numerous incidents clearly illustrating ferrocement's good resistance to heavy impact. Severe impact, particularly point impact, may of course cause damage, but it tends to be localised, and can later be easily and quickly repaired. My own experience of this is a good example. During a recent trip down the Thames in a ferrocement yacht, we collided with a tug, the edge of its transom digging into the hull. The force of the collision was considerable, but fortunately the damage was slight, and localised. It is highly likely that if the yacht had been constructed in GRP or wood, the damage would have been more severe, and more extensive in area. The collision area was easily repaired, and quickly too. Fragments of mortar clinging to the mesh were retained, all loose particles brushed away, and the hole replastered with a batch of mortar, leaving the hull as strong as before.

Other vessels have collided with submerged objects, jetties, other craft and mooring buoys, and their owners have all reported either no damage at all, or damage for which only minor repairs were required.

The tensile strength of ferrocement can be high, assuming an average figure of about 5,340 lbs/sq. in. ultimate. The comparative tensile strength of high quality timber can reach about 6,000 lbs/sq. in., but it must be remembered that this is only with the grain; while across it, tensile strength, from a structural consideration, is almost negligible. Tensile strength of a ferrocement hull is in all directions, by virtue of the mesh reinforcement. Where compressive strength is concerned, a ferrocement hull can achieve an ultimate strength of about 12,000 lbs/sq. in. in 28 days of curing, which figure will continue to increase with age, far in excess of that which can be obtained with wood.

Ferrocement hulls are fire resistant. Wood and GRP will burn or melt, while a steel hull tends to buckle and distort. Only very high temperatures will damage ferrocement. One reason is that the constituent materials, mortar and steel rod, have similar coefficients of expansion, which means that they will expand and contract at the same rate with changes of temperature. There is therefore no, or very little, cracking or separation at the steel/mortar interface.

Many people fallaciously believe that a concrete hull would be damp and cold; a belief obviously conjured up from their own experience with land-based concrete structures. This, though, is not a characteristic of ferrocement. The material has a very low thermal conductivity. This being the case the rate of heat transference is low, thus allowing for the probability of a ferro hull remaining cool in hot weather (a distinct advantage in the tropics, particularly if the deck and coach roof are of ferrocement) and relatively warm in cold climates. The material is of course impermeable, so there is no dampness, and there shouldn't be any leaks either. I have also noticed that there is very little condensation in a ferrocement hull.

Another characteristic of a ferrocement hull is the space saving that can be achieved, in relation to other methods. The average skin thickness of, say, a 35-footer would be in the region of $\frac{3}{4}$ in. By the very nature of the structure, and also depending on the design of the hull, fewer beams or frames not in the shell would be required to give the structure strength and rigidity. In a wooden boat of similar size and design, the hull thickness would be probably twice that of the ferrocement boat, and would also require far more ribs. Space saving in the concrete boat could amount to as much as 8 to 10 per cent, which can make quite a difference to the comfort below.

Ferrocement is still regarded as a relatively heavy building material, which in general is unsuitable for the construction of small craft under about 18 ft in length, or for those under about 35 ft designed with light displacement. Nevertheless a number of small boats have been built in this medium. The Dutch scow mentioned earlier was little more than a dinghy, and the first ever ferrocement boat was only about 12 ft over all. This weight disability disappears as size increases. The turning point is around 35 ft to 40 ft, where concrete boats can and have been built as light as their wooden or steel counterparts. The length/weight ratio improves considerably above this, when ferrocement craft can be built much lighter than similar sized wooden and steel vessels, and as light as those constructed in GRP.

There is, as yet, no known upper size limit for ferrocement craft. To my knowledge the largest yacht yet built in this medium is a 73 ft ocean racer. Other larger sea-going vessels have been built, while I believe a barge some 350 ft over all is successfully in operation. Using present practices and design criteria, vessels up to 100 ft over all could well be constructed. Above this length it is likely that the stresses exerted on the hull would warrant a more rigorous design analysis than has yet been provided.

At the present time ferrocement clearly offers numerous possibilities and advantages over the more conventional materials. Developed as far as it now is, the technique is more suitable to the amateur builder rather than

4 One of the largest ferrocement yachts ever built in Britain, the 65 ft *Petralis*, being fitted out at Tower Pier, on the Thames in London. The owner is New Zealander John Aston who completed her construction in November 1970.

5 *Rolling Stone*, a 32 ft ferrocement sloop berthed at Spring Lake, Michigan, June 1968. This vessel is one of the early examples of ferrocement yacht construction. Together with a sister vessel, she was built at Kalamazoo, Michigan in 1934 by builders L. W. Young, H. S. Adams and J. L Dane. According to her owners, she is left in the water all the year round, and during winter has been frozen into from four to five feet of ice with no damage occurring. The first time she was ever pulled out of the water for any repair was in 1961. All that required doing was sandblasting, patching up a few small areas on the hull and painting. The yacht has been taken all over the Great Lakes, spent much cruising time near Panama, and despite a few hard knocks has come through with flying colours. The owner reports only marginal trouble occurring with the ferrocement deck due to small cracks forming. In the winter water would get in, freeze and enlarge the cracks. A good layer of fibreglass on the decks remedied this problem. Photo: M. E. Iorns, Fibersteel Co.

to commercial production line scale. Low cost materials, the elimination of any necessary formwork or moulds, and negligible maintenance requirements are undeniably attractive characteristics.

But the 'convert' should not be too eager to rush off straight away to order his truckload of cement and steel. I have pointed out some of the advantages of this medium and, perhaps, at first sight the technique may appear simple and straightforward, well within the capabilities of any amateur shipwright. This is not strictly so.

One should understand at the outset that success can only be achieved if certain important considerations are taken into account. The design limits of ferrocement are much narrower than for some other building techniques. There is less latitude for error, and what errors there may be will be less easy to rectify. Very rigid selection and quality controls of component materials must be exercised if a sound and satisfactory hull is to be completed. For example not just any type of mesh can be successfully employed; not every type of cement is suitable; nor any type of aggregate. Mix design is critical, more so than with other concrete structures. Proper curing is also vital.

And probably to most of you who may be contemplating building in ferrocement the material and its constituents are entirely new. Experience and information on the technique is not yet that widely spread, so it could be that you are, in your own right, a pioneer of the medium in your locality, unless you happen to be building in an area where other concrete boats may have been constructed. There is the possibility that there is no one to turn to for advice when you strike a problem, in which case you will have to rely on your own capabilities and initiative.

What I am driving at is that while ferrocement techniques are described as suitable for the home builder, there are home builders and home builders. *Yachting Monthly* magazine put it well: 'Home building is a term covering anything from a glue-and-screwdriver man to an expert working in his own time. The individual must define himself to himself.'

Numerous papers and articles have covered fairly fully various aspects of ferrocement boat construction, from its early development to the practical application of the technique. Two books have been written on the subject which, in general terms, cover the technique reasonably well. Nevertheless it has become increasingly apparent to me that there is need for a further manual, and it is my hope that this book will contribute toward resolving this need.

A layman myself, I am writing for laymen, particularly those prospective amateur builders who desire a simple working guide or step-by-step instruction on how to manipulate the ferrocement technique so that a sound and satisfactory end product can be obtained. Remember that there is nothing more worthwhile before embarking on a project than adequate preparation.

ferrocement 2

FERROCEMENT DEVELOPMENT

1855-1943
1943-1971

Despite the fact that ferrocement has only recently climbed into recognition as a suitable construction technique for boats, it is most definitely not the product of current technological development. As a modern construction technique it dates back to the 1940s when it was re-introduced and defined by the Italian engineer Prof. Pier Luigi Nervi. To review its origins, however, we have to go back some 120 years to the 1840s, when its introduction to the world heralded the advent of steel reinforced concrete as a structural material; and, in fact, it is believed that its first application was in the construction of a boat.

To understand why ferrocement or steel reinforced concrete evolved it is necessary to understand something of the limitations of concrete as a building material. As we understand it today, concrete is composed of a hardened matrix, formed by the chemical reaction between cement and water, and an aggregate, usually sand, to which the hardened cement adheres. The quality of the concrete is largely governed by the careful selection of the constituent materials, the proportions in which they are combined, and finally how the hardening process is controlled.

As a building material concrete fulfills an admirable role as a mortar or glue to bind together stone, gravel or most other materials. Used on its own as a structural material concrete has a limited role, however. In most structural applications there are considerable tension stresses which have to be accommodated. And, unfortunately, while concrete has tremendous compressive strength, it is very weak in tension. On its own it would be unable to withstand these stresses.

It is undoubtedly this lack of strength in tension which prompted the development of reinforced concrete. By imbedding steel in the concrete the structural qualities of the composite were greatly enhanced, with the steel improving its bending, shear and torsional strength considerably. The concrete, from being brittle and completely unreliable when used on its own, was completely transformed.

This idea of reinforcing concrete did not emerge suddenly like a spark in the dark. It was more a gradual and continuous process of development. Centuries ago attempts had been made to improve the structural strength of concrete. Many different materials have been used, even today, but none, thus far, have proved to be as satisfactory as steel. Bamboo and straw have been among the materials used. Reinforcement has also been attempted externally. It wasn't until the middle of the nineteenth century that anyone hit on the idea of imbedding steel in concrete to reinforce it. Metal had been used, by the Romans first, I believe, as an external reinforcement in the form of bands or straps. Even when the idea was first introduced it wasn't fully understood. According to the historical records, the man responsible for first developing this concept was the French horticulturist Jean-Louis Lambot, from Mirival in Provence. His very first project was a rowing boat. His technique, while being understandably rather crude, closely resembles the principles adopted by today's ferrocement builders.

Lambot was born in 1815 in the town of Montfort, quite close to Mirival in the south of France. The son of a wealthy landowner who had large estates at Mirival, Lambot spent much of his early adult life in Paris, where he studied law. He also travelled a great deal, and it is quite possible it was during this period that his imagination was stimulated to formulate many of the ideas which he attempted to develop later in life. Apparently, he had no real interest in pursuing a career in the profession he had studied for, because it was not long before he returned home to take over the management of the family's estates.

It was there that he carried out his pioneering work with mortar and steel. It seems that his initial experiments were directed to such objects as plant pots, water jars and other receptacles; probably for various applications on his estates. Years later the remains of such artifacts were discovered at Mirival. Perhaps the results he obtained with these simple items—ease of construction, durability, impermeability—prompted Lambot to experiment with more ambitious projects.

In 1848 or 1849, at the age of 35, he hit on the idea of replacing a wooden boat on the lake on his estate with another constructed with mortar and a form of iron reinforcement. The technique Lambot adopted he later labelled *Ferciment*. The world's first concrete boat was not large. It measured about 12 ft in length and had a beam of about 4 ft. The hull was quite thin with a cross-sectional thickness of not much more than an inch. The construction consisted of a mesh or grid reinforcement made up of two layers of small diameter iron bars set at right angles to one another. This structure was then plastered with a cement mortar, which only thinly covered the reinforcement. A narrow seat, constructed in the same manner as the hull, surrounded the interior, and a metal capping strip formed a protection for the gunwales.

The basic construction technique adopted by Lambot is obviously not identical to that used by today's ferrocement builders. It is similar in principle, however, and is substantially different from conventional reinforced concrete practices. The size of this first boat somewhat confounds the often-expressed view that ferrocement, because of its weight, is an unsuitable material for small craft.

Lambot's experiments with ferrocement did not stop with that one boat. Soon after, he completed another which was, perhaps, slightly more sophisticated in construction. At any rate it appeared to stand up better to years of use and exposure to the elements than did the prototype. Both craft were recovered from the bottom of the Mirival lake in 1955; while both were in reasonable shape, the second version was definitely in better condition.

Convinced of the potential of the new construction technique he had developed, Lambot decided to offer it to the public. In 1855 he placed his first boat on show at the Paris Exposition. The exhibit was stuck away between 'a small boat with a screw having moving blades suitable for steam navigation' and a 'machine for towing boats'. Its description in the official catalogue was brief:

04094 Lambot (J-L), a Carces (Var)—Bateau

Accompanying the boat was a short description of the method of construction, a list of advantages the material was supposed to have, together with a few of Lambot's ideas for other possible applications. These included swimming pools, barges, buoys, water cisterns and cellar doors. The advantages Lambot listed are substantially the same as those which Nervi mentions for the more sophisticated ferrocement technique he developed 100 years later. They included: a saving in initial cost of construction; saving in maintenance (this is negligible); speed of construction; immediate repairs in case of damage; impermeability; soundness under test.

The year after the Paris Expo Lambot successfully took out Belgian and

French patents for *Ferciment.* At the same time he applied for a British patent. The title given to the specification read: 'An improved material to be used as a substitute for wood in naval architecture and other constructions, and also for improved domestic purposes where dampness is to be avoided.'

Here, it is needful to mention that Lambot was not the only person at this time experimenting with steel reinforced concrete. His application for a British patent was preceded by a year by another presented by an ex-plasterer from Newcastle-on-Tyne, W. D. Wilkinson. Wilkinson's techniques differed from Lambot's *Ferciment* in that they resembled more closely those adopted in conventional reinforced concrete structures, in which the concrete predominates rather than the steel reinforcement. Numerous other 'pioneers' were handing in patents for reinforced concrete, but for the purposes of this exercise these other developments can be largely ignored, for the techniques conformed more closely to conventional reinforced concrete practices than to ferrocement; and, at that stage their applications were strictly for land-based structures.

Lambot's unique method of boatbuilding was almost totally ignored at the time. The attitude was at best non-interest and at worse disbelief. It is this reaction, not only from the public sector, but in specialist architectural and constructional fields which has persisted almost to this day, and has been a major reason for the slow development of ferrocement. It obviously blocked Lambot's enthusiasm, deterring him from developing further his technique. There appears to be no record of any other ferrocement structures he may have built. Instead he apparently turned his attentions more to the running of his estates. In later years he became known more for his knowledge in matters concerning agriculture than as the inventor of a unique construction technique. Lambot died in 1887, aged 73.

It is quite likely that Lambot's contribution to the development of reinforced concrete, and more specifically ferrocement bôatbuilding, would never have been recalled but for the efforts of a Marseilles building contractor, M. Gassier Bastide. In 1902, more than 50 years after the construction of the Lambot prototype boat, Bastide, while on a visit to the Mirival estates, came across one of the craft still floating on the lake.

He immediately recognised the boat's construction for what it was. He also noted that the concrete did not have the raw, new appearance of the concrete structures of those days, and realised that it must have been fairly old. Around the estate Bastide also found further artifacts in mortar: the pots and jars. Bastide brought his find to the notice of the technical world, but it was another 50 years before *Ferciment* and the Mirival boats gained any sort of official recognition.

In 1949 a centenary celebration was held in Paris to commemorate reinforced concrete. The conference was presided over by Vincent Auriol, President of the French Republic. It was here that the first official tribute was paid to Jean-Louis Lambot's pioneering efforts. As a result of one of the debates a search was initiated to recover some relic of the Frenchman's work. Six years later the search proved more than successful. After draining the lake at Mirival, a careful search in the silt and mud of the bottom uncovered the two original rowboats. One of them, referred to earlier, was still in remarkably good condition; a good example of the durability of ferrocement. Unfortunately this boat has since been lost. But the other, undoubtedly the prototype, is still very much in existence today and holds pride of place in the museum at Brignoles, not far from the Mirival estates where it was built more than 120 years ago.

1855-1943

During this period of almost 100 years there were few further projects which can in anyway be associated with ferrocement development. Those there were contributed little to the further understanding of the medium, or its development. Certainly none related to naval architecture. Only one case warrants a mention here, and that only because it does provide further evidence of the remarkable durability and hardiness of ferrocement.

Coincidently, the year Lambot died, in 1887, a Dutch construction company built a scow following ferrocement principles. It happened to be the first of a series of experiments they conducted with reinforced concrete. The boat was called *Zeemeeuw* and closely resembled, structurally, those built by Lambot, in which a grid framework was set up and plastered with mortar. As a variation it had a simulated clinker hull. Whether or not the company took their lead from the Frenchman is not certain. Quite possibly the technique they adopted was purely the product of their own research and development. Anyway, it was very similar to the earlier *Ferciment*. What is of interest about the *Zeemeeuw* is that, up to three years ago, it was still in good condition and still in use on the pelican pond at Amsterdam Zoo. It has recently been placed in a museum. More than 80 years old and still in use is certainly a good testimony of the lasting qualities of ferrocement.

Just for the record, there was a great spate of reinforced concrete shipbuilding at this time. This activity was concentrated mainly in the years 1917–22 and during the later stages of World War II. The construction techniques adopted followed conventional reinforced concrete practices. The large-scale destruction of shipping during the hostilities brought about pressures for an accelerated shipbuilding programme. This position was made worse by the acute shortages of plate steel. Reinforced concrete provided a suitable alternative to the usual methods of construction. It was much cheaper for a start, and was an ideal material for mass production. During the earlier period some 150,000 deadweight tons of seagoing shipping was built, not counting several hundred barges, pontoons and lighters. About 77 concrete vessels were built in Britain during World War II.

It is quite possible that ferrocement techniques, as developed by Lambot and followed up by others, were contemplated during these periods instead of ordinary reinforced concrete methods. But the material wasn't sufficiently understood at that time to do much with it. Even the knowledge on conventional reinforced concrete practices was extremely limited. And the urgency of the building programme did not allow any time to carry out research.

The results were predictable. Design limits were not known and the authorities decided that they were going to play safe. The ships built were grossly over-designed and consequently 'over-weight'. No attempt had been made to provide hull shapes which would be more in keeping with the properties of the material. Instead they were merely copies of steel vessels of equivalent size. Most of these ships performed their emergency roles admirably nonetheless. But it is, perhaps, understandable that further production during peacetime was discontinued. They were far too heavy to compete adequately with other commercial vessels. They were forced to carry much ligher or smaller cargoes than their equivalent-sized counterparts, but at the same operating costs. This obviously made them an uneconomic proposition.

1943–1971: Period of Redevelopment

Louis Lambot is acknowledged as being the founder or inventor of ferrocement. However, it is to Prof. Pier Luigi Nervi, an Italian engineer and co-founder of the well known firm of Nervi & Bartoli, based in Rome, that most of the accolades must go for the development of this material. For it wasn't until the 1940s, when Nervi re-introduced the material in a more sophisticated form, and defined it, that the characteristics of the medium became more clearly understood. It is unlikely that Lambot was fully aware of the function of the steel reinforcement in mortar or of the properties of ferrocement as he knew it.

Prof. Nervi's name has been linked for a long time with the development of reinforced concrete structures: he is a person much respected in the architectural and civil engineering worlds. To the layman he can be associated with contemporary structures which have received wide acclaim, such as the Exhibition Hall in Turin, and the designs for the buildings housing the Rome Olympics. In many of them ferrocement techniques were adopted during the construction.

His first experiments with ferrocement began in 1943, and his account of this early work has been well documented. Perhaps the most widely read paper is that which was published in the first number of the 1951 issue of the Italian publication *L'Ingegnere*, under the title 'Ferrocement: its characteristics and potentialities'. This paper, written by Prof. Nervi, was translated into English by the Cement and Concrete Association. His conclusions form much of the basis of present improved practice.

Nervi's development evolved from the fundamental concept behind reinforced concrete: that concrete can withstand large strains in the neighbourhood of the reinforcement, the magnitude of the strains depending on the distribution and subdivision of the reinforcement throughout the concrete. Extending this concept was the observation that the elasticity of a reinforced concrete member 'increases in proportion to the distribution and sub-division of the reinforcement throughout the mass'. Nervi's initial experiments developed from this theme and in structure they closely resembled the *Fercíment* invented by Lambot almost 100 years previously.

Nervi built a series of thin mortar slabs reinforced with mesh formed by the tying together of mild steel wire of 0.5–1.5 mm diameter into a grid with squares roughly 1 sq. cm in area. The mortar had a rich cement content: something in the region of 60 lbs of cement per cu. ft of sand. Various concentrations of reinforcement (more or fewer layers of mesh. the gauge being held constant) were experimented with, and in some experiments mild steel bars were sandwiched in the middle of the slabs, the thickness of which ranged from 6 cm to 10 cm.

The resultant material, Nervi noted, 'had very little in common with normal reinforced concrete'. It possessed the 'mechanical characteristics of a completely homogeneous material', which combined the freedom of form of concrete and the strength of steel. Nervi chose to describe the method of construction as 'ferrocemento'. The composite exhibited two important qualities: extreme elasticity and a great resistance to cracking, which, together with its elimination of the need for formwork (thus reducing costs) and potential as a mass production method, suggested to Nervi that ferrocement might have an application in naval architecture.

He was able to interest the Italian naval authorities in the material to the extent that late in 1943 a feasibility study was embarked upon. This involved a series of more detailed and accurate experiments than had been previously carried out, under the supervision of the Italian Naval Register. The work was done in two parts, by Nervi's construction firm Nervi & Bartoli at Viareggio (southern Italy), and in Rome by Prof. Martenelli at the laboratories of the Experimental Institute for Building Construction.

The experiments were designed to determine the mechanical characteristics of ferrocement and the design limitations. One of the experiments was to subject ferrocement panels to impact tests to determine its resistance to collision shock. The panels were about 5 sq. ft in area and about $1\frac{3}{4}$ in. thick. A 550 lbs weight was dropped from different heights, eventually increasing to almost 10 ft. When damage or failure eventually occurred there was no actual penetration of the slab. No complete hole was formed. Instead the mortar crumbled away, and the mesh was deformed and weakened. But what was important was that the damaged area still showed some vestige of cohesion, with particles of mortar still adhering to the mesh reinforcement. An obvious advantage, particularly in naval structures, for the particles would impede to some degree the passage of water.

The qualities exhibited by the ferrocement during the tests persuaded the Italian Naval Register and the Department of Marine Engineering to accept in principle ferrocement as a suitable construction technique. As a result, 1943 was not yet over before Nervi & Bartoli received orders for four vessels to be built in this material. The first of these was a 400 tons displacement motor vessel, which was built in a combination of precast reinforced concrete frame units and a ferrocement hull. The other three vessels were somewhat smaller in size being only 150 tons deadweight. In their construction even the frames were of ferrocement. However, none of these vessels were ever finished beyond the hull stage. According to Nervi, the military events at the time caused the projects to be abandoned.

Two years later, soon after the war ended, Nervi & Bartoli again took up the ferrocement cause. They had an Italian architect, Guido Franco, design a 165 tons displacement motor yacht which they built in the yard of Lazzarini & Meacci at Anzio. Christened *Irene*, the hull of the yacht was only $1\frac{3}{8}$ in. thick, constructed out of three layers of 0.24 in. mild steel bars, with two of the layers placed longitudinally as stringers and the other vertically. All the bars were at about 4 in. centres. Sandwiched round them were eight layers of mesh, four on the inside and four outside. The 'skeleton' was finally wired tightly together. A rich mortar was used, pozzolanic cement being the principle constituent. This was mixed to the consistency of thick paste and forced through the reinforcement from the inside, by hand, and trowelled smooth on the outside. The vessel was completed in three months. No formwork was used as the mesh structure of the reinforcement acted more or less as laths to retain the mortar. Unfortunately Nervi does not make any reference to the type of frames used to initially hold up the reinforcement and provide the hull form. However, in later projects steel pipe was used, so perhaps similar materials were adopted for the construction of *Irene*.

Nervi recalls that the total weight of the vessel was 'approximately five per cent less than that of an equivalent vessel in timber', while the construction costs were almost half. The concrete hull proved to be a complete success; entirely watertight with no evidence of the mortar being permeable. After five years of use no maintenance was required, except where a couple of accidents had caused some damage. In each case, colliding with the sea bottom and striking a submerged wreck, the damage was not serious and was localised more or less to the point of impact. And as was exhibited in the earlier impact tests, the crushed motar in the damaged area tended to cling to the reinforcement, impeding the flow of water into the vessel. In each instance repairs were easily effected: cement mortar was clapped on the damaged section, sealing the leak.

Following the building of these four vessels another crisis struck the small shipbuilding industry. Capital was tight and subsidies for the construction of any vessels, whether they were in wood, steel or ferrocement, were severely restricted. Nevertheless, 1946 saw further work being carried out in ferrocement. This included a 20 ton floating crane and a fishing trawler. Nervi also entered the realm of yacht construction, and that same year built the 41 ft ketch *Nennele*. Apparently the yacht is still in good condition and in use today.

In building this vessel Nervi made use of steel tubing to provide the hull form. This method is almost universal today. Pipes with an internal diameter of 25 mm (1 in.) were bent to the hull shape at designated stations at 1 m (3 ft) centres. Longitudinal rods, 6 mm in diameter (0.24 in.), were stretched round the formwork at 5 cm (2 in.) centres. The whole was then sandwiched between seven layers of mesh, bringing the cross-sectional thickness of the hull to about 12 mm ($\frac{1}{2}$ in.). Ballast for the boat was placed inside the trough keel. Nervi reported that some of the yacht's apparent advantages over similar sized vessels constructed in wood were cheapness of construction, greater interior space, and the absence of maintenance costs for the hull.

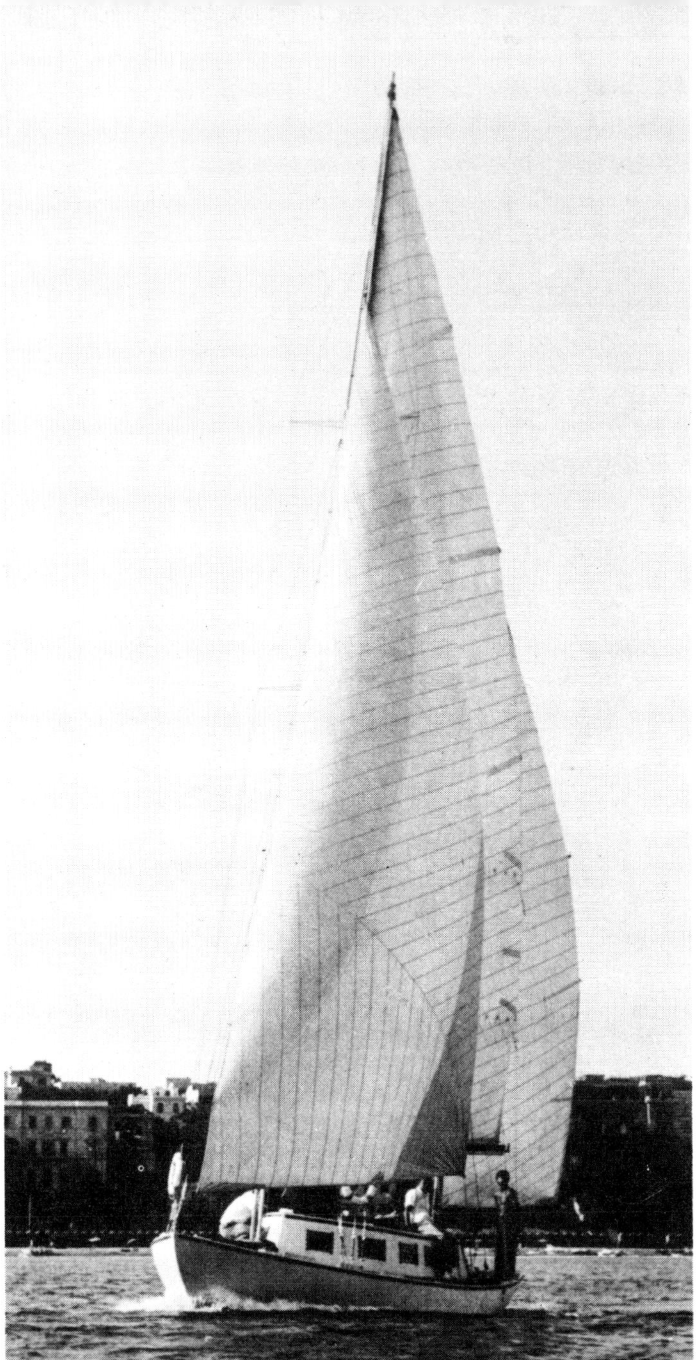

6 The yawl *Nenelle* designed and built by Prof. Pier Luigi Nervi during the early 1940s. Despite several hard knocks she is still in sound condition after more than 20 years of use.

At about the time Nervi was building his yacht another Italian, Prof. Oberti, was carrying out further laboratory experiments with ferrocement structures at the Polytechnic Laboratory in Milan. He came up with some revealing results which greatly increased the information about the material, particularly relating to the elasticity of ferrocement. One of the most enlightening results to come out of the experiments was the importance of the steel content in relation to the volume of concrete. Oberti's results were further substantiated during the 1960s when L. D. G. Collen carried out similar experiments in Ireland.

Oberti's test methods involved measuring the degree of extension or elongation of a ferrocement member at the point where cracking occurred. Various trial concentrations of reinforcement were experimented with in order to determine which proportions of reinforcement and mortar were best. At the lower end of the scale, where the proportion of steel was least, and in Oberti's test slab measured between $7\frac{1}{2}$ to 15 lbs per cu. ft, it was found that the elongation of the composite resembled that of unreinforced mortar. Which meant, of course, that there was very little elasticity and the tensile strength was minimal.

However, as the steel content was increased so the structural and mechanical qualities of the composite were seen to improve. Prof. Oberti eventually reached an optimum limit in steel content in which the measurement was 30 to 40 lbs per cu. ft. Here, he noted that elongation became almost five times as great as that exhibited by the ferrocement slab with low steel content. It was also observed, not only by Oberti but by Nervi as well, that a fairly constant balance occurred between the weight of steel and the amount of mortar when a dense mesh was impregnated and only just covered by the mortar. In practice this weight is estimated to be in the region of 24–36 lbs per cu. ft. To go beyond this density would not substantially improve the extensibility of the composite to make up for the greatly increased weight. A further problem would be the difficulty of achieving complete penetration of mortar through the mesh matrix.

Despite some very obvious advantages, Nervi's 'up-dated' technique was not received with much enthusiasm from the boatbuilding fraternity. It took roughly 15 years for them to build up any interest. But the initial reaction was sufficient to cause Prof. Nervi to abandon his efforts to further stimulate the application of ferrocement to marine construction. Until quite recently, that is, when his firm became involved with the construction of two fishing vessels for the FAO.

It is almost impossible to pinpoint the reasons which stimulated the sudden build-up of interest in ferrocement at the beginning of the last

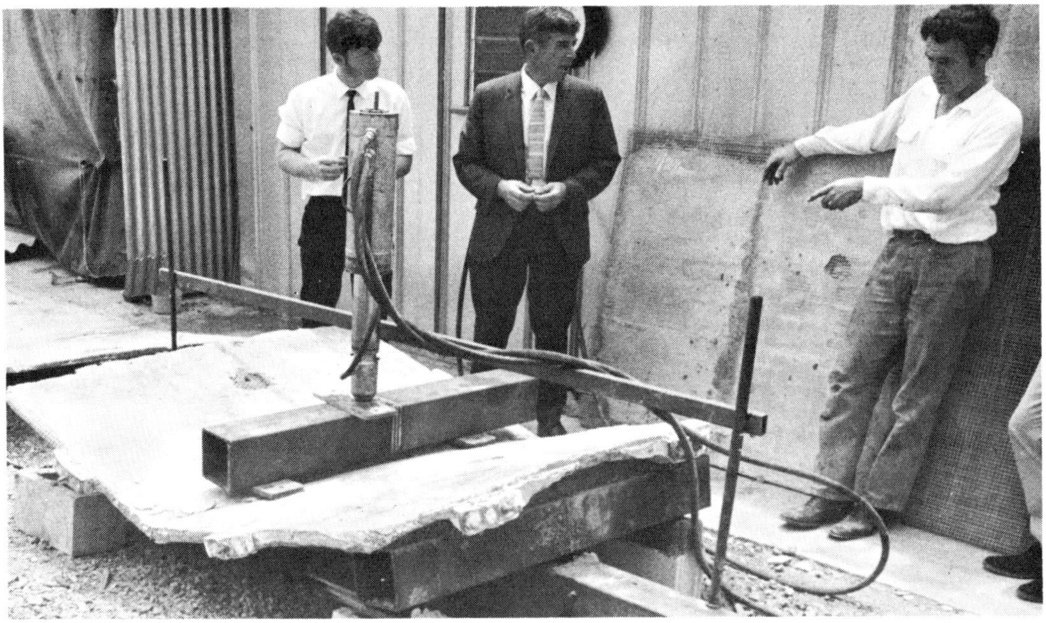

7a Bending and shear tests on a sample ferrocement panel by Australian firm Elstress Ltd, near Sydney. Tests in this case were on a 5 ft span of 0.8 in. minimum thickness ribbed panel. Illustrated is a panel bent to its failure point after cyclic loading. Load at failure was estimated at 13,400 lbs, deflection at failure was 8 in., elastic deflection with a 95 per cent recovery was 1.2 in., equivalent stress at failure was 9,300 lbs per sq. in.

7b Impact tests being carried out on a sample ferrocement panel at Elstress (Pty) Ltd. The panel was 0.8 in. thick and a 15 lb iron ball was dropped from a height of 5 ft, with the area of impact failure measured over 10 blows. These tests were being carried out prior to the construction of a 71 ft racer in ferrocement.

decade. Undoubtedly they must have been partly economic. Then perhaps there is now a greater familiarity with reinforced cement structures which has resulted in a more sympathetic and less biased attitude toward the material. Whatever the reasons, they created a climate which has obviously favoured the further development of ferrocement techniques and their application to marine construction.

Those boatbuilders who were in the van of this 'revival', which shows no signs of being of limited duration as on previous occasions, relied almost entirely on the techniques developed by Prof. Nervi (using a pipe framework to provide the hull form) and the work carried out by Prof. Oberti regarding reinforcement proportions etc. Theirs was the definitive work, and even now the techniques have not differed significantly. Of course there is now a much deeper understanding of the medium and there have been a number of improvements.

The experimental work carried out by L. D. G. Collen, member of the Institute of Civil Engineers (Ireland) at Trinity College, over 10 years ago further added to the knowledge of ferrocement by providing some definite values for the mechanical properties of the material. The conclusions Collen arrived at as a result of his experiments back up those of Nervi and Oberti. Like Oberti, Collen constructed a series of ferrocement beams which were similar in dimensions to the structures used in practice. The experiments were planned to determine bending strength, shear strength, the modulus of elasticity and also variations of these with differing water/cement and cement/aggregate ratios. The sand used in the mix was a good unwashed pit sand which was finely sieved, 80 per cent passing through a 14 sieve (BS). The cement selected was rapid hardening Irish Portland Cement, while the mesh used for reinforcement was an 18 gauge mild steel woven mesh weighing approximately 1.4 lbs per sq. yard. From 4 to 12 layers of mesh were used, sandwiching $\frac{1}{4}$ in. mild steel rods laid at 3 in. centres. Steel content varied from 17 lbs to 32 lbs per cu. ft.

Collen observed that there was a direct relationship between the ultimate bending stress and the steel content, with its capacity to withstand such stresses improving at a constant ratio in relation to increases in reinforcement. Likewise there was a constant gain in shear strength of the ferrocement unit as the steel content increased. Modulus of elasticity of the unit escalated rapidly with slight increases of steel content up to about 30 lbs per cu. ft. Relating the curves of all three mechanical qualities it was determined that a steel content of around 24 to 36 lbs per cu. ft contributed the best design values. Below are some of the figures drawn up by Collen:

NOTE! consistent w/ P. Nervi (see p. 36)

Young's modulus of elasticity: 820,000 to 1,300,000 lbs/sq. in.

Bending stress (working load): 1,000 to 1,200 lbs/sq. in.
Shear stress (working load): 100 to 400 lbs/sq. in.

These design figures are a considerable improvement on those for conventional boatbuilding materials. The test data produced larger values.

Other information Collen was able to obtain from his experiments related to the ratios of cement/sand and water/cement which exhibited the best properties for ferrocement. Using a constant water/cement ratio of 0.35, which figure, incidently, Nervi used in his structures, Collen varied the cement/sand ratio. It was noted that as the ratio increased so there was a rise in bending strength, reaching an optimum at a ratio of 0.66. After this there appears to be a slight drop in strength.

The water content of a mortar is obviously very critical. While an increase will bring about better workability of the mix it will undoubtedly cause a drop in strength, and also increases the rates of shrinkage and creep. Collen reported that with the water/cement content of a mix increased from 0.30 to 0.35 there was a drop in strength of 350 lbs per sq. in. Correlating the two ratios linearly on a graph the conclusion was that the figure of 0.35 was a reasonable value for the water/cement ratio, to provide adequate strength of the mortar and yet allow for sufficient workability. This ratio is generally adopted by most ferrocement builders today. And in practice the ratio of cement to aggregate is in the region of 1:2, though it is sometimes richer in cement. These mixing ratios can be calculated as volume/volume or weight/weight.

Since Collen's report was published in 1960 other data has become available concerning the physical properties of ferrocement. The information is fairly consistent, and does substantiate previous results. That from the British firm Windboats Ltd is related to their Seacrete patent and was included in a test report prepared by Lloyd's Register of Shipping. Density of the ferrocement is 151 lbs/cu. ft, while ultimate tensile strength is 5,340 lbs/sq. in.; modulus of elasticity is 1.30×10^{6} lbs/sq. in; compression strength of the mortar (after 28 days) is 12,225 lbs/sq. in. Thermal conductivity of the ferro was about one sixth that of steel.

These test results by Lloyd's for Seacrete were further confirmed by a study of ferrocement conducted in 1968 in the Department of Naval Architecture of the University of Michigan. This study was carried out by Hans Muhlert, a graduate student at the university, with two associates, J. F. Coleman and N. Jergovich. The subject worked on was a 108 ft scalloper designed by John Gilbert Associates of Boston. This was originally designed for wood construction, for which comprehensive design data is available. It was selected in order to make a comparative study with

8 28 ft TSDY motor cruiser built to Lloyd's 100 A1 classification by Windboats Ltd, Wroxham, Norfolk. The vessel has a ferro-cement hull built to the Seacrete patent.

an identical counterpart built in ferrocement. Experimental strips of ferrocement composite, conforming to Nervi's original specifications, 1$\frac{3}{8}$ in. to $\frac{3}{4}$ in. thick, were made up and subjected to tension testing to determine actual bending strength. Besides verifying results obtained by other researchers, the object was to secure more precise values for accurately determining scantlings for ship design. Significant of Muhlert's results was that they apparently indicated an upper limit of 1,200 lbs per sq. in. allowable stress for the calculation of ship design. It is noted, however, that the material tested contained an extremely high steel content of 50.8 lbs per cu. ft. In practice this steel content is far beyond the capabilities of the ordinary boatbuilder, who can barely reach 25 lbs per cu. ft, the recommended minimum concentration for ferrocement.

ferrocement 3

DESIGNING FOR FERROCEMENT

Introduction
Various designs

Introduction

Once the major decisions have been taken to build a boat and to build her in ferrocement, the next priority is obviously the selection of a design in line with your particular requirements and within the limits of your planned financial commitment. As ferrocement is to be the construction technique, we can safely assume in the first instance that the boat is not going to be dinghy-sized.

Generally the minimum practical size of boat which can be built in this medium is around 18–20 ft. Below this the weight disadvantage becomes too great, and there are other materials far more suitable. There is no known upper size limit for ferrocement. A 150 ft seagoing barge is being built in the US, and some of Nervi's successful craft were over 100 ft long. However, one structural engineer has suggested that 100 ft length overall may be the present limit taking into account the stage of present engineering design knowledge. Beyond this size differences in stresses occur and a more rigorous stress analysis would be required than most pleasure boatbuilders are yet capable of handling.

With regard to capital outlay and the proposed intention of building the largest sized vessel which can be managed, be not misled by the apparent cheapness of ferrocement construction techniques. True, there can be a saving of up to 50 per cent on the hull construction. But remember that the cost of the hull will perhaps contribute only about a third of the total lost of the completed vessel. And while the hull costs may not substantially increase as size goes up, the cost of equipping the vessel and fitting her out will undoubtedly escalate rapidly.

Obviously the selection of a good and pleasing design is of paramount importance. If the idea is to draw up a design yourself, you would do well to think very carefully about it first. Generally the basic requirements of a sea-going vessel are seaworthiness, comfort and speed, in that order, coupled with ease of handling, especially if at any time she is to be sailed short-handed. But to achieve a balanced combination of all these facets is not easy or always successful, even for the most experienced naval architect. And it is highly unlikely that an amateur designer will have that degree of skill and art to produce a good design. Note the many hideous examples floating around of the amateur designer's work.

To spend a lot of time and money building a vessel which may be structurally sound, but disappointing in performance and probably aesthetically monstrous, is foolhardy. The job is best left in the hands of those who have spent years acquiring the expertise to handle it. There are now numerous designs available for the prospective ferrocement practitioner. Some of them are published with this chapter. More and more naval architects are coming to terms with the medium, realising that there is a growing demand for designs drawn up specially for ferrocement.

It is not necessary to limit your selection to the offerings of the various advertised ferrocement plan services. If one of their designs fills your requirements by all means buy it, but at least attempt to assess how successful the architect has been in meeting his specification. Provided that you avoid multihulls or light displacement monohulls below 35 ft l.o.a. there is a wide selection of stock plans available, which were originally drawn up for other construction techniques, and have been proved as successful designs. However, it must be realised that these designs will have to be modified to accept the new construction technique. One should understand that every vessel is designed for the materials which are going to be used in her construction. By adhering to a few basic rules any good moderate or heavy displacement hull, previously designed for, say, timber or steel construction, can be converted for ferrocement.

Contrary to the opinions expressed by many so-called ferrocement pundits, ferrocement boats can be built to practically any conceivable shape, provided that the reinforcing can be formed to it. In this I make reference to the statement that ferrocement cannot be used for flat sections as, for example, are found on single chine, double chine and multi-chine hulls or flat bottomed barges. It can most definitely be used for these configurations but, just as with GRP, these sections require special stiffening and ribbing to eliminate flexing. To do this properly requires an intimate engineering knowledge of the material. The hull form which would make the most efficient use of ferrocement is that which has good compound curves. By virtue of this shape all external forces exerted on the hull are converted into compression loads which will be carried by the reinforcement within the ferrocement shell.

This can be explained further by reference to the principles of membrane theory and shell theory relating to thin-walled structures, which is what a ferrocement hull is. The theory suggests that the curvature of a load bearing member is such that the principal stresses are placed in the plane of the surface. These requirements would therefore be met by a well-curved hull. The limit to this concept is that a flat membrane is unable to support any lateral loads. In the presence of such forces it would bend or deflect and stretch in an attempt to develop the necessary curvature. This means that a hull with flat surfaces (chines, multi-chines, flat bottoms) would be subject to flexing in these areas unless they are suitably strengthened with frames or ribbing. The same situation would apply to other flat surfaces, like the deck, floors or bulkheads.

At the present stage of ferrocement development there is no established statement or specification of precise values which can be used to accurately determine the strength of ferrocement for specific thicknesses or scantlings for ship design. True, the research work of Collens, Byrne and Wright, and Nervi has to some extent opened the door to the mechanical properties of the material, but far more empirical research must be done to establish criteria for calculating the structural design of ferrocement forms.

To date most ferrocement hulls have been designed without conclusive strength estimates being carried out, or at best these estimates have largely been guesswork. Many architects have tended to base their scantlings design on previously built ferrocement vessels which have proved to be sound. The usual method of obtaining data reflecting the load-carrying capacity of specific ferrocement composites is by experiment with representative test panels. But it must be realised that only rough design estimates can be determined in this way. References to these were made in the last chapter.

Recently Hans Muhlert, during an extensive study of ferrocement (I mentioned this project briefly at the end of the previous chapter), appears to have made a breakthrough in this area. Muhlert's study revolved round the mathematical analysis of ferrocement, the results of which (according to a report in the American journal *National Fisherman*) have 'permitted the development of a method of accurately predicting the load-carrying capacity of the material'. In this specific study, which made use of design data for a 108 ft wooden built scalloper, it became possible to

9 *Suisun Clipper*, $54 \times 17 \times 6$ Pacific type salmon troller and combination boat designed by Edwin Monk. Hull, deck, soles and bulkheads are in ferrocement. Production mould and prototype built by Fibersteel Co., West Sacramento, for Carus Marine Industries, Suisin City, Calif.

calculate scantling dimensions for a 'theoretical ferrocement vessel equivalent to the scalloper'.

Another member of the University of Michigan, Charles D. Canby, briefly describes the method Muhlert evolved in a paper on ferrocement which he presented to the Pacific Northwest section of SNAME in March 1969. Muhlert's work, incidently, was written up in a report in 1968 entitled 'Ferrocement Trawler, Design Study Report'. Co-authors with him were N. Jergovich and J. F. Coleman. Canby describes Muhlert's approach to structural design as the 'transformed area concept', a method which is covered in most books dealing with reinforced concrete design. He says: 'The transformed area concept uses the fact that the unit stress in steel reinforcement is n times as great as the stress in the concrete. Therefore, 1 sq. in. of steel may be considered equivalent to n sq. in. of concrete. In applying this method, all rods in a beam are assumed to be replaced by an equivalent number of square inches of concrete in the same location with regard to the neutral axis.' The concept can be represented thus:

$$n = \frac{E_s}{E_c} = \frac{\text{Modulus of elasticity of steel in compression}}{\text{Modulus of elasticity of concrete in compression}}$$

Muhlert has since produced a more significant report of the structural design of ferrocement members. Published by the University of Michigan in Jan. 1970 (No. 043), it is entitled 'An Analysis of Ferrocement in Bending'

To explore further the concept of designing for ferrocement construction I asked Maurice Griffiths to give me some general comments, which he has kindly done on the following pages. To the British yachting fraternity Maurice Griffiths, GM, needs no introduction. His name has been intimately linked with the sport and the industry in this country for more than half a century. Mr Griffiths was Editor of *Yachting Monthly* from the end of 1926 until 1967 when he retired. He is the author of a number of books on yacht cruising, design and construction. He also has a considerable reputation as a designer, specialising mainly in shoal draught centreboard and bilge keel types, as well as a number of ocean cruising yachts. Altogether some 800 yachts from 24 ft upwards have been built to his designs in all parts of the world. More recently he has been turning his hand to designing in ferro-cement, and modifying some of his already well-tried designs, previously drawn up for traditional timber construction, for the new medium. Two of his modified designs, which originated from his 38 ft ketch called *Good Hope* and a class known as the Tidewater, are reproduced with this chapter. Following are Mr Griffiths' comments.

A ferro hull, at its present stage of development, could be described as a strong flexible structure of steel pipes, rods and closely knit steel mesh which is made impervious by impregnating and coating this fabric with a plastic cement mix.

Although there are many wooden built yachts and commercial vessels like fishing boats, harbour launches and workboats which can be reproduced in ferrocement, or ferrocrete as it is sometimes called, certain modifications have to be made to the hull so as to convert satisfactorily from wood construction to the more plastic material. The designer should therefore bear in mind both the methods of construction to be followed and the limitations imposed by them. In fact he can almost forget much of the

10 Maurice Griffiths designed this Good Hope class shallow draught yacht, shown plastered. The boat was built by Peter Brooks at Penpol Boatyard. The yacht was constructed using conventional water pipe frames and the mesh used was 22 gauge $\frac{1}{2}$ in. weldmesh. Note that the decks have also been fabricated in ferrocement. Photo: R. S. Foster.

technique of wooden hull construction, for the methods of ferro boatbuilding owe much more to the practice of steel construction. He would do well first to study small boatbuilding in steel if he is to bring the right mental approach to the problems of building in ferrocement.

This material offers almost endless possibilities, and because in its present form ferrocement boatbuilding is still in a development stage (25 years or so is accounted as newfangled in most pleasure boating circles!) new builders are constantly bringing fresh ideas and techniques to the subject. Indeed, I for one should not be entirely surprised if in time a lighter and more resilient material will be found for the plastering in place of cement. Or the steel reinforcement replaced by an equally tough but much lighter material.

In today's ferro hulls the main structure is normally formed of mild steel pipe frames of $\frac{1}{2}$ in., $\frac{3}{4}$ in. or 1 in. internal diameter, according to the size of the vessel, which are spaced at regular intervals from bow to stern like the timbers or frames of a wooden ship. If these pipe frames are continued down and under the hull each side so as to form a hollow trough keel, as is normal practice with steel and in many GRP hulls, the meeting of the sides of the keel and the vessel's bottom (at what in a wooden yacht would be the garboard strakes) must not form an acute angle. The bottom should blend into the keel sides in a gentle curve to which the pipe frames can satisfactorily be bent without buckling. The same will apply to the bottom of the keel if the tubing is to be bent round twice to form the trough: the keel bottom must not be formed with too sharp curves.

At the deck edge also (the covering board in a wooden vessel), if the deck is to be of ferro and the pipe frames continued round the hull like large hoops so as to form part of the deck, the curve between the topsides and the deck must again not be sharper than the minimum curve to which the pipes can be bent. The radius of this curve can generally be taken as not less than five times the inside diameter of the tubing. Furthermore, not only is it difficult to bend steel pipe without heating into very sharp curves, but it is almost impossible for a builder to spread the necessary six or eight layers of wire mesh evenly round a sharp bend or a corner without ruining the shape and appearance of his boat.

Single or double-chine hulls intended for plywood planking are not satisfactory for building in ferro (though they can be built if necessary), even if the design is modified and the chines rounded off somewhat. The rounding off of chines and garboards still leaves sharp curves that make it difficult for the builder to shape his pipe frames and wire his layers of mesh evenly. Another factor against the adoption of a chine or V-bottom design for building in ferro is that there must still be large flat areas over the topsides and bottom. Whereas these flat areas are suitable for plywood with its adequate strength in all direction, a large flat surface in a ferro hull will tend to flex under stress—as when bashing into a steep sea or striking the ground in a swell.

Like an egg the shell of a ferro boat is strongest if it is well rounded everywhere. In designing a boat specially for building in ferro, therefore, the garboards should be given sweet curves and topsides sufficient curve in flare and tumblehome so as to avoid unnecessarily large areas of flatness. In this way the immense strength inherent in this material will be fully utilised.

In some craft pipe frames are cut off at deck level, separate angle iron deck beams are welded to the pipe ends, and pipe of greater diameter (say $\frac{3}{4}$ in. or 1 in.) is run like a beam shelf from stem to stern, and welded or wired at every pipe frame/deck beam angle. With this method the deck edge can be almost square as in a wooden yacht. Otherwise, the topsides can be continued above deck level to form a rail or bulwark, being pierced or interrupted where necessary to form deck scuppers.

Although the hollow trough, filled as requisite with iron scrap aggregate, is the simplest and most common form of ballast keel for ferro yachts, in some projects a separate concrete/iron scrap keel has been cast on the site in prepared shuttering made from the architect's drawings. With this type of keel the bottom ends of the pipe frames are welded in pairs to athwartships flange plates or webs which have been precast in the keel. Provided these webs or plate floors are well embedded in the concrete and the welded joints to the pipe frames securely made, I do not think trouble should arise from corrosion and weakening of the joints in years to come from this method of construction.

For a very shallow draught yacht, like a centreboarder, this precast form of keel has one advantage over the trough, namely that the bottom corners need not be radiused; in fact the section can be square. With this shape every inch in the depth of the keel sides within the limited draught will be gained for lateral resistance when sailing to windward.

For those designers and owners accustomed only to wooden yachts and fishing vessels, with their massive oak or laminated stems and stern posts and keels and floors—built like a church as they used to say of many a boat in the past—there might be a temptation to overdo the weight of the ferrocement boat's framework. There is no need, for instance, to use a piece of heavy section channel or a girder for the stem or stern frame, nor $\frac{1}{2}$ in. or $\frac{3}{4}$ in. solid bars for the framing or deck structure. All this weight is something which should be avoided.

Ordinary steel water or steam piping of light gauge for stem, keel base loop, decks and carlines, when knitted to a closely intersected fabric of steel reinforcing rods and welded wire mesh, makes a very strong homogeneous structure, and need never be oversize or overweight. Although small ferro boats between 25 ft and 40 ft in length are usually, but not invariably, somewhat heavier than their wooden sister ships, most ferro vessels above 40 ft start to turn the scales, and the 50 ft and 60 ft yacht can be given a somewhat better ballast ratio than some of its orthodox wooden competitors.

It is clear that no two builders are likely to make up their keel in identical ways. One with his own welding equipment may decide to strengthen his trough keel at intervals with steel plates welded at the edges of the pipe frames; another, perhaps a non-welder, will divide the trough into compartments by making his webs in ferro in the same manner as the rest of the hull. A third might not use the trough keel at all, but rather bolt onto steel floor frames an existing cast iron keel bought cheap from an old yacht recently broken up, or he may cast his concrete keel on the site as described above.

Whether you use steel angle plates or ferro webs for your keel it is a simple matter to bolt wood cross bearers to the tops of them for supporting the cabin sole or floorboards. Likewise the engine bearers and fuel and water tank supports (if the tanks are not moulded integral with the hull) can be built-up ferro mouldings with angle plates or hardwood cappings bolted to them for ease of attachment and trimming.

Whether to design with all-ferro deck and deck structures, or a ferro deck with wood coamings and deck beams and plywood cabin top, or wood deck beams, beam shelves, carlines, coamings and a ply deck throughout over an otherwise ferro hull depends, generally speaking, on the size of the yacht and on the importance of weight saving at deck level, and on how much the owner wants his ferro yacht to look like a traditional wooden one. Very roughly, in normal cruising yachts below 27 ft waterline laid or plywood decks with wooden coamings and cabin top are worth the saving in weight over ferro. But above 35 ft waterline the weight saving advantage of wooden decks and structures is hardly perceptible, and only the aesthetic aspect in the owner's view justifies the added expense of a wooden lid.

How much ballast should a normal ferro yacht need, and what bulk of concrete/iron scrap aggregate would this make? Also, would a mixture of lead scrap and concrete be heavier and therefore give better ballast/weight ratio? Good questions, and worth fair answers. The total weight of ballast required in the keel must be calculated by the designer as for any wooden, steel or other type of yacht; but he must bear in mind that if the decks in a small yacht (say 30 ft or under) are also of ferro the extra weight here will slightly raise the position of the yacht's centre of gravity, and more ballast will accordingly be needed to give her the same stability or sail carrying power. Just how much more must be calculated by the designer since the amount will vary with every boat.

Cast iron alone varies in quality from one foundry to another and can weigh anything from about 436 lbs to 460 lbs per cu. ft, making roughly 5 cu. ft to one ton of 2240 lb. Lead, on the other hand, is fairly constant at 720 lb per cu. ft, or say 3.1 cu. ft per ton (English or metric). An average coarse sand/cement mix goes roughly 140 lbs cu. ft, and the weight of scrap iron aggregate will naturally depend on how much scrap you have been able to work into the mix. For all practical purposes a rule-of-thumb guide can be taken as 240 lbs per cu. ft, or say 9 cu. ft to the English ton. This may appear somewhat bulky compared with a cast iron ballast keel, but not to worry if the boat you are planning is to be a full-bodied cruising yacht with ample space in the trough keel beneath the floorboards.

Despite the attractions of lead with its high density for yacht keels, lead is not a wise choice for a ferro boat. In seawater lead and iron or steel are uneasy bedmates, for when they are in contact or very close a violent electrolytic action is set up, and the ferrous metal—the iron or steel—is rapidly wasted away. Unless the pieces of lead can be securely insulated from the steel parts of the hull and the bilgewater it is as inadvisable to put lead in a ferro hull as it would be in a steel hull.

NOTE!

Before the yacht is launched and tried under way it pays the builder to leave a compartment or two in the trough keel empty both forward and aft of the ballast aggregate, so that if it is found that she needs to be trimmed a little by the head or stern, trim can be adjusted simply by pouring more ballast aggregate into the appropriate compartments. To ensure that no empty compartments forward of the ballast are unable to drain aft into the pump suction well (which should be accessibly placed for attention to the pump filter or strum box near the yacht's heel) some builders have laid a covered duct or 1 in. diameter pipe connecting the keel compartments before pouring the ballast mix. All compartments must have some form of limber hole to allow the bilgewater to drain back to the pump.

Lightening ↓

It is happily not common for us mortals to be struck by lightning, for rough justice from the gods acts usually in a less dramatic way, but a number of yachts and workboats have been struck during violent storms. In steel boats the electrical discharge usually finds a path down one of the shrouds or stays into the hull from where it passes fully earthed into the surrounding sea. A wooden boat, on the other hand, can have its masthead destroyed and various metal fastenings in the hull blasted or burnt out, unless the masthead has been connected to a stay which is itself bonded to a metal strip led directly to one of the keel bolts or some other kind of earth plate in direct contact with the water.

The ferro hull on the other hand offers unique conditions. Because it is in effect a fabric of small diameter steel rods, piping and layers of comparatively fine wire mesh encased in cement, any intense electrical discharge could spread through a large portion of the hull, and if there was no clear path to a good earth, the result would be widespread fusing with extensive damage to the hull, or even a mild explosion with partial destruction of the vessel. Especially in parts of the world where violent thunderstorms are prevalent it behoves the builder of a ferro vessel to protect her from lightning strike and its consequences.

This need not be difficult with a yacht if one or other of the mast stays is treated as a conductor to an earthing plate in contact with the water. One way of making a lightning conductor which has been used, and which I have incorporated in ferro designs of my own, is to bond the masthead ironwork to the foretopmast stay which in turn is bonded to the stemhead fitting and the main stem pipe which leads down to the keel loop. At some convenient spot below the waterline an earthing plate is welded to the stem pipe with the outside face of this plate in direct contact with the sea. By this means the electric discharge has a clear path from the masthead down the stay to the stemhead and stem to the earth plate from which it should be dissipated into the water.

If the yacht has a bowsprit the bonded stay carries the current through the cranse iron (bowsprit end fitting) via the bobstay to the stem pipe and thence to the earth plate. Alternatively a heavy gauge metal strip can be bonded to the lower end of the bobstay and so led down under the water. A similar conductor, if preferred, could be led via the masthead backstays to metal strips at the yacht's quarters which are led down below the waterline.

Although in attaching shroud plates in a ferro vessel there is a temptation to make use of the construction and embed the plates like the pipe frames and wire mesh inside the cement mortar so as to make a very neat job both inside and out, this is not wise practice. Moulded in shroud plates cannot be got at, and you have no way of knowing if they are beginning to corrode and weaken, until one lets go in a breeze and you perhaps lose your mast overboard.

Furthermore, should you find it advisable to move the mast, or alter the yacht's rig or layout of the shrouds, you cannot alter the positions of these built-in shroud plates. It is better, therefore, to attach the shroud plates with bolts in the same way as on a wooden hull, the plates being made to fit the ship's side. Mechanically it is best to fit shroud plates on the outside of the hull, although racing practice has created a fashion for inside plates because they leave smooth topsides. In a ferro hull outside plates enable you to fit corresponding plates on the inside of the hull to which the shroud plates are bolted, making a secure and sound fastening.

To illustrate how a well-tried design for traditional wood construction may be modified so as to be successfully built in ferro, two of my cruising yacht designs are reproduced. The first is a 38 ft ketch with centre cockpit and stern cabin which was designed for a family to build in ferrocement in Australia for a voyage round the world. The origin of this design was a 38 ft ketch called *Good Hope*, which I had designed specially for a Suffolk family to be stoutly built in trad timber fashion by Whisstock's Boatyard at Woodbridge.

She was planned for blue water cruising, but her draught was limited to 4 ft 3 in. for exploring Continental waterways. The *Good Hope* and two sister ships subsequently built in wood proved fine healthy dry sea boats, easy for youngsters to steer and manage their comparatively small sails, and this appeared to be a design that appealed to a large number of other offshore-minded families.

Today's high costs of traditional construction made an alternative design, suitably modified for home-building in ferro, an attractive proposition for many prospective builders. The design shown was the result. To allow for an estimated increase in overall weight of the ferrocement hull over its wooden sister, plus the greater bulk required for the concrete/iron scrap trough keel, the underwater profile was deepened a little and a draught of 5 ft 3 in. aimed at. The turn at the garboards of the wooden *Good Hope* was eased and the hull slightly modified as required for simple methods of ferro construction.

This is on a framework of $\frac{1}{2}$ in. galvanised steel pipe frames spaced every 24 in. from stem to stern, looped around the stem and keel and decks to the edge of the coachroof coamings. Solid $\frac{1}{4}$ in. diameter steel rods are spaced over 3 in. centres both fore and aft and athwartships and wired together, and eight layers of $\frac{1}{2}$ in. square 19 gauge welded steel mesh complete the reinforcement. While the decks, like the hull, are of ferrocement, the coamings, coachroof and hatches of main and after cabins are traditionally built in mahogany.

Under the ketch rig shown these yachts sail themselves on course with wheel becketted (or emergency tiller shipped and lashed or shock cord held) with the wind anywhere from close-hauled to well abaft the beam. For lone voyaging one of the vane steering gears could be fitted, when it might be simpler to rig the yacht as a Bermuda cutter. The 10 in. wide plank bowsprit would carry the masthead jib with a Wykeham-Martin jib furling gear for working from the cockpit, and twin running staysails could be installed if required for the trade wind routes.

The second design has been built as the Tidewater class in wood in various parts of the world and numbers 16 boats to date. Originally a shallow draught cruising boat with a 2 ton lead ballast keel and bilge keels drawing 3 ft 4 in. when upright, the Tidewater has proved to be a very tough, capable little sea boat with many hard winds up to Force 10 to her credit. Although only 30 ft over all (excluding the plank bowsprit) she has 9.5 ft beam and a surprising amount of room below decks. A ferrocement version of this simple type of cruiser was called for, and the design shown was the outcome.

With raised foredeck and trunk coachroof with short doghouse and ample side decks, the deck plan owes something to those fine modern cockle bawleys plying out of Leigh Creek in Essex. But the keel has been deepened to accommodate the weight of aggregate ballast required in the trough, and there are no bilge keels required, for the draught is 4 ft 8 in. With a quarter berth the layout will sleep five, and there is space in the fore cabin for an additional Root berth to port or starboard for a sixth if necessary. The doghouse extension over the fore end of the roomy cockpit gives both crew and helmsman shelter on a hard thrash to windward, and like the 38 ft ketch, this boat, whether Bermuda or gaff cutter rigged, will sail herself with helm pegged, or lie-to in strong winds, for hours unattended.

Construction is similar to that specified for the 38-footer but with $\frac{1}{2}$ in. diameter pipe frames spaced every 24 in., $\frac{1}{4}$ in. steel rods fore and aft and athwartships at $2\frac{1}{2}$ in. spacing, and six layers of $\frac{1}{2}$ in. square welded steel wire mesh over $\frac{3}{4}$ in. stem pipes and keel loop.

Other designs

For contrast, here are a few other designs drawn up specifically for ferrocement construction. Understand that all the plans published are the copyright of their respective designers and should therefore not be copied or used for the building of any craft without the permission of the designer.

The first three plans which follow Mr Griffiths' Good Hope and Tidewater designs are representative of those stock plans carried by Ferro Cement Marine Services, a small company which is based at Burnham-on-Crouch in Essex. The company is led by the young boatbuilder John Hodsdon, who has, as his resident architect, designer Christopher Petrie. The company is prepared to design boats from 25 ft to 75 ft l.o.a. and has a number of stock designs available up to 44 ft l.o.a. They are also prepared to build hulls or kits to the customer's design, or supply the labour for any particular stage of construction which may be required. Their kits comprise frames, stem, keel, sternpost and horn all bent to shape and marked out ready for assembly, together with full construction drawings.

The Fibersteel Co., West Sacramento, California, was formed in 1960 by Martin E. Iorns and is one of the leading concerns specialising in ferrocement marine construction. Fibersteel pioneered the cavity mould process, which incorporates a female mould similar to that used in glass fibre production methods, and appears to offer advantages for commercial series production.

Martin Iorns commissioned American designer Jack Rouse to prepare a set of plans for a 32 ft and a 55 ft hull which could be used for series production using Fibersteel's ferrocement construction techniques. The smaller hull has two sailplans and four powerboat designs for the same basic hull configuration. The 55 ft hull, Valeo (from the Latin for 'strength'), is available as a motor yacht or motor-sailer. In the development of the Valeo design Captain Irving Johnson of *Yankee* fame provided advice and assistance, and it is clear that he much influenced the shape of the end product.

As a motor yacht the Valeo will carry an engine which should develop over 100 h.p. Cummins Diesel Co., after an analysis of the hull configuration, have indicated that their 175 h.p. engine would drive the boat at 10.3 knots maximum with a cruising speed of 9.6 knots. The motor-sailer version of the Valeo, which is offered structurally complete for 12,000 dollars, is fitted with twin centreboards for better lateral resistance, facilitating her trim for off the wind sailing and ease of handling. She also has a wide flat keel for grounding which is heavily reinforced. Essential dimensions of the sailing version of the Valeo are: l.o.a. 55 ft, l.w.l. 45 ft, beam 15 ft 6 in., draught (board up) 4 ft, (board down) 8 ft, displacement 26.2 tons, total sail area 1122 sq. ft. The hull thickness is $\frac{3}{4}$ in. minimum tapering to 6 in. at the keel.

The dimensions of the 32 ft hull are waterline 27 ft, beam 11 ft, draught 3 ft.

11 *Adonde*, $55 \times 15\frac{1}{2} \times 4\frac{1}{2}$ Valeo class ferrocement motor-sailer launched in 1970. Fibersteel Co. has priced the structurally complete boat—hull, deck, cabin, interior bulkheads, soles, floors and tanks—at $12,000.

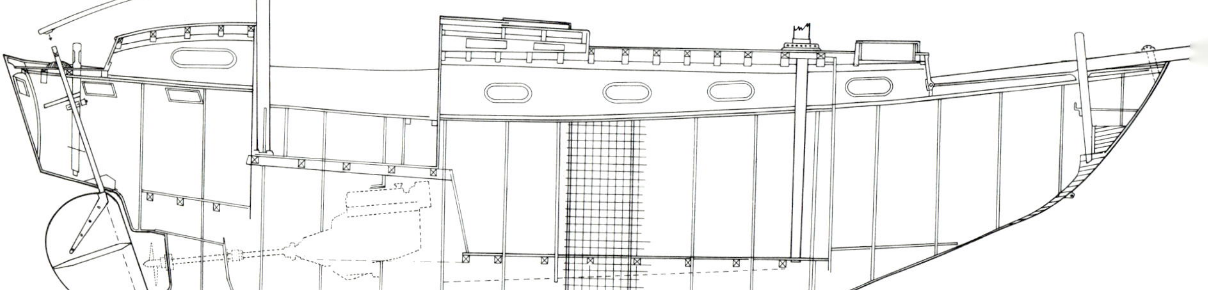

Good Hope Class		
LOA	38 ft 4 in.	10.16 m
LWL	31 ft 2 in.	9.50 m
Beam	11 ft 0 in.	3.35 m
Draught	5 ft 3 in.	1.60 m
Displacement	28,500 lbs	9,500 kg
Ballast	9,000 lbs	4,200 kg
Working sail area: ketch	707 sq. ft	66.0 sq. m

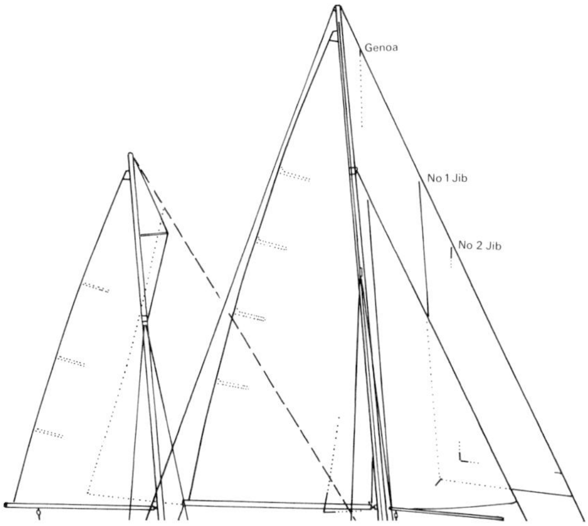

1

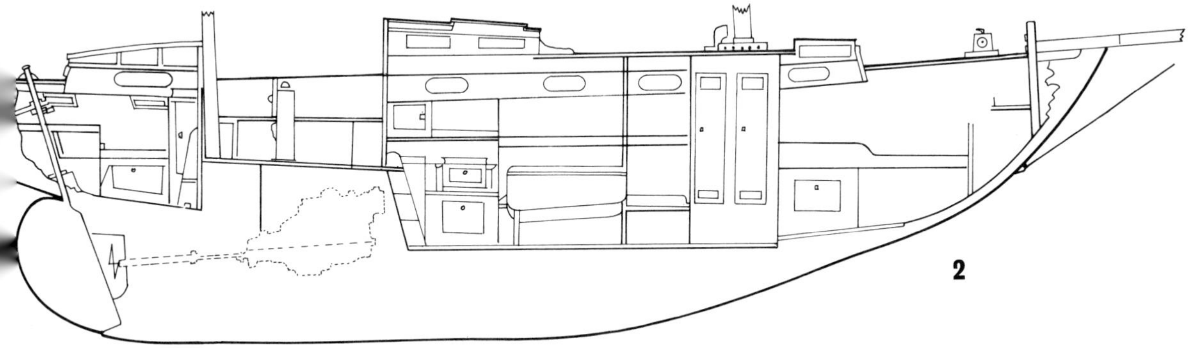

2

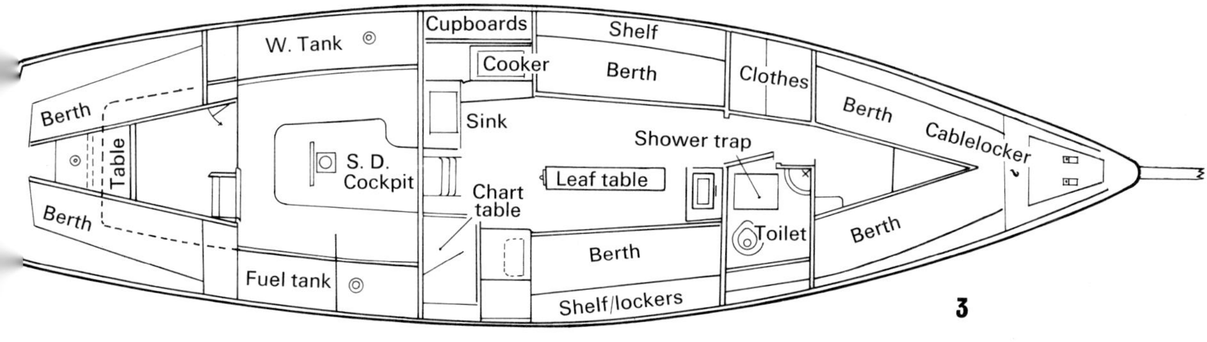

3

1, 2, 3 Good Hope class 38 ft ketch, designed by Maurice Griffiths

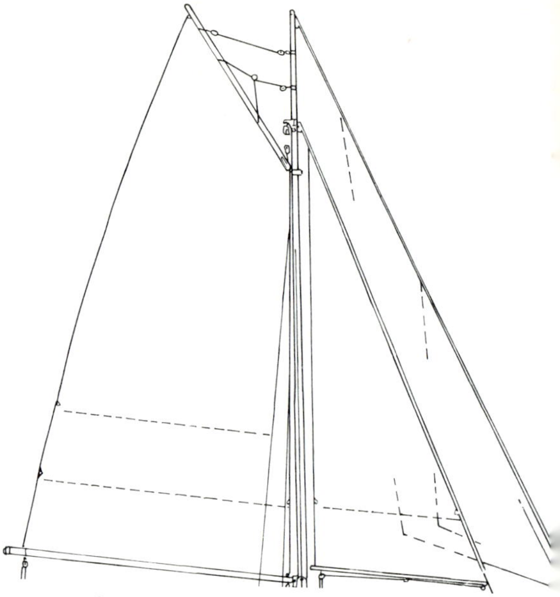

Tidewater Class

LOA	30 ft 0 in.	9.1 m
LWL	26 ft 2 in.	7.98 m
Beam	9 ft 6 in.	2.90 m
Draught	4 ft 8 in.	1.42 m
Displacement	14,400 lbs	6,368 k
Ballast	5,760 lbs	2,630 k
Working sail area		
bermuda cutter rig	498 sq. ft	46.0 sq. m
gaff cutter rig	514 sq. ft	47.2 sq. m

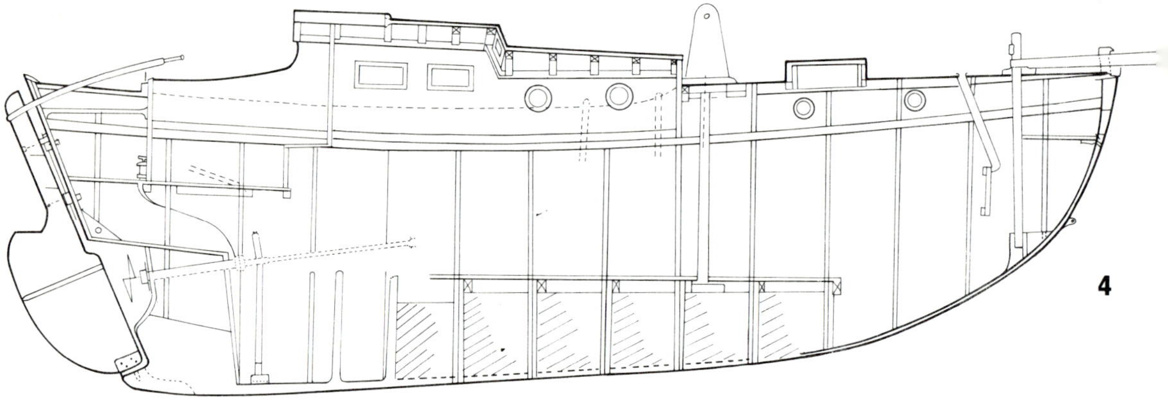

4

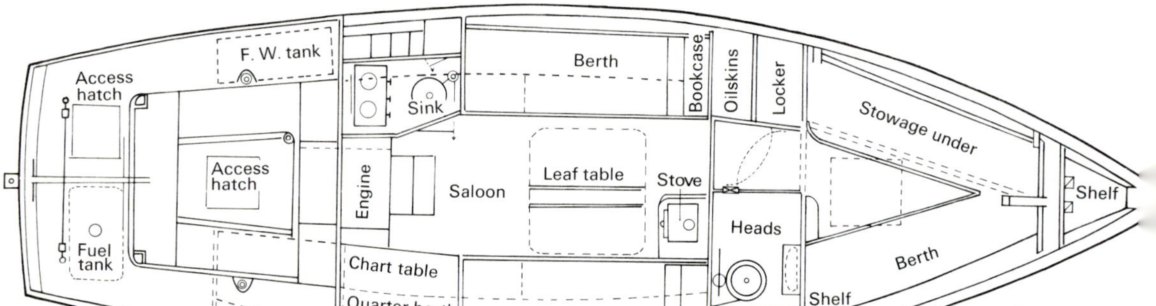

5

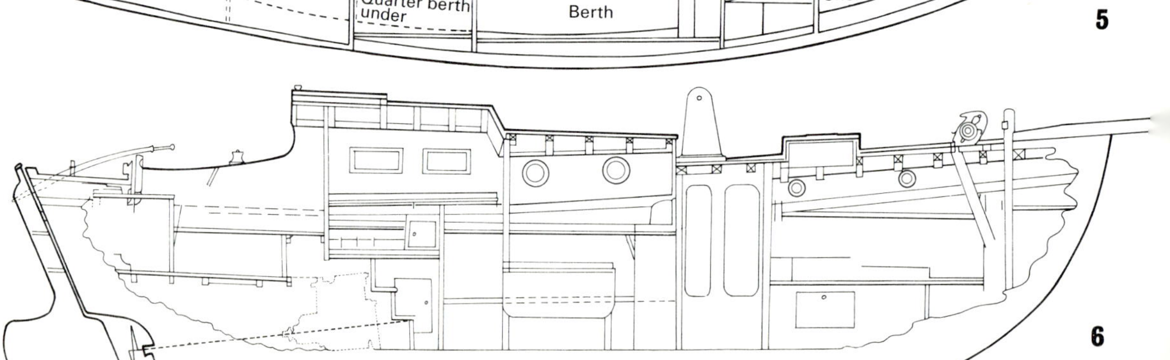

6

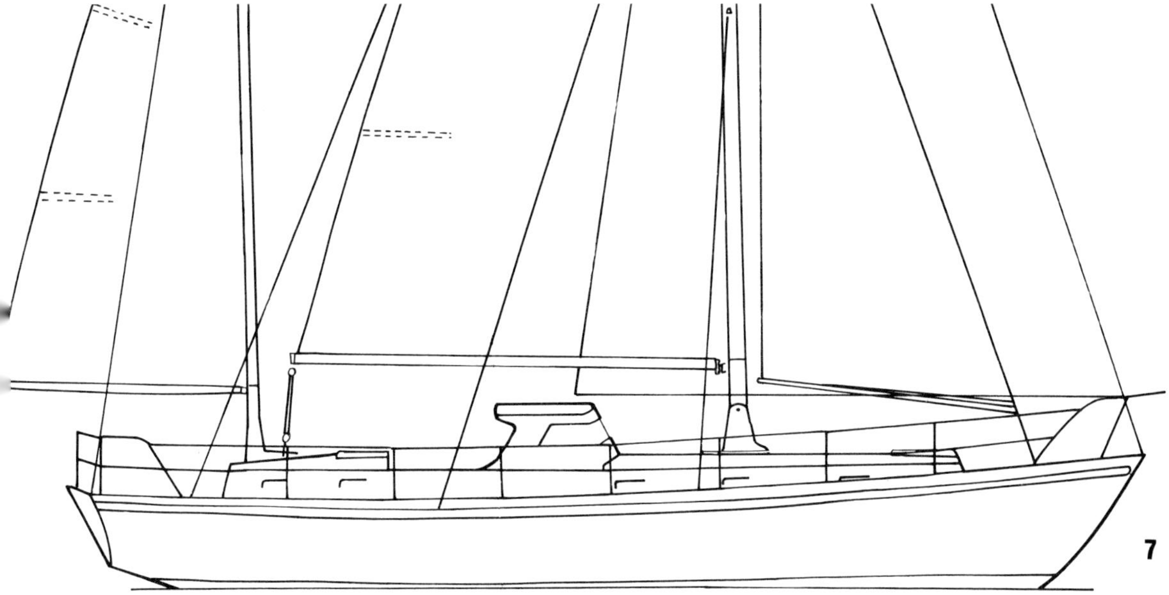

7

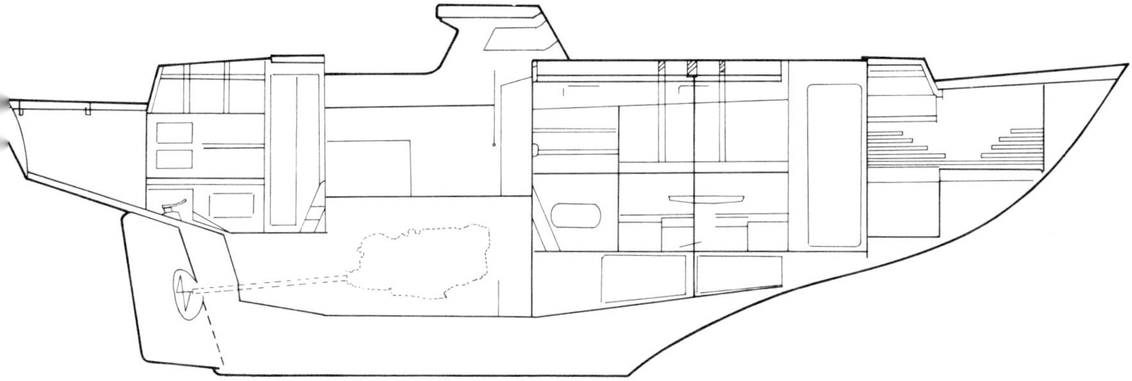

8

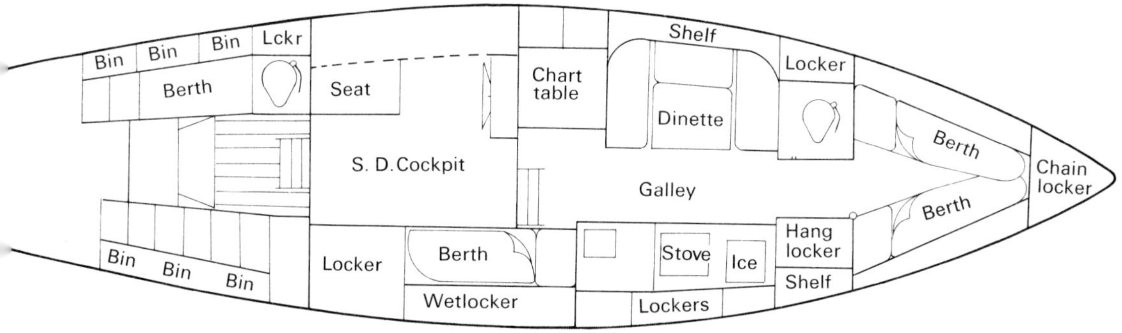

9

4, 5, 6 Tidewater class 30 ft cutter, designed by Maurice Griffiths

7, 8, 9 Plans for centre cockpit 40 ft ketch, designed by C. Petrie for Ferro Cement Marine Services

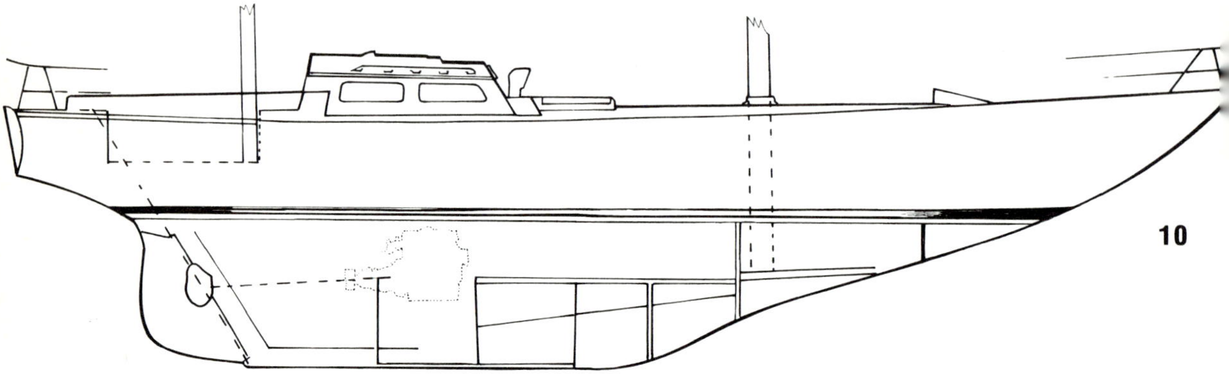

10

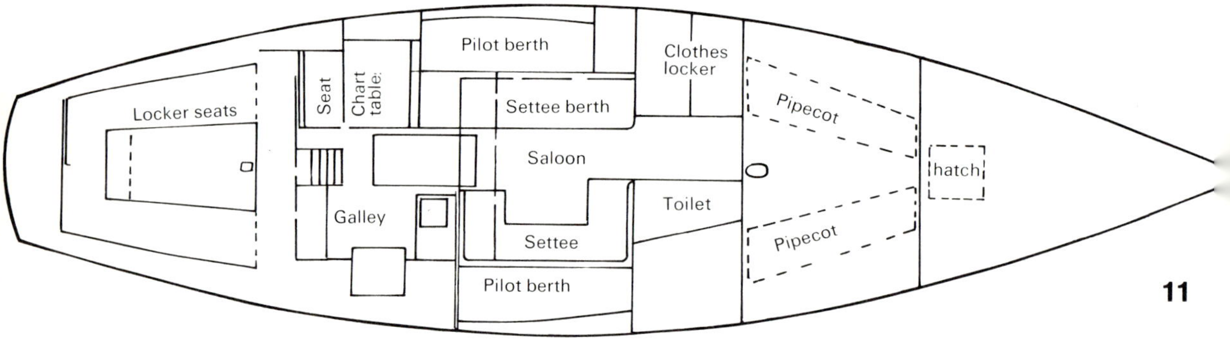

11

10, 11 Plans for 46 ft ketch, designed by Alan Hill and under construction by Ferro Cement Marine Services

12, 13, 14 Plans for 55 ft ketch designed by C. Petrie

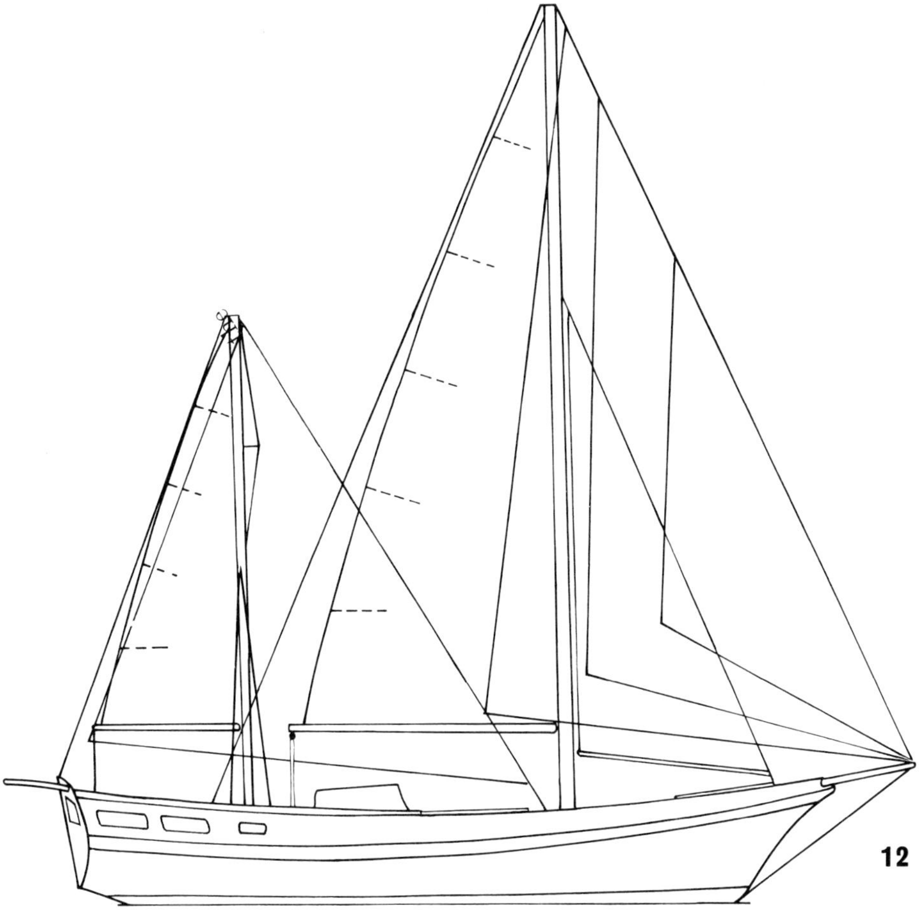

12

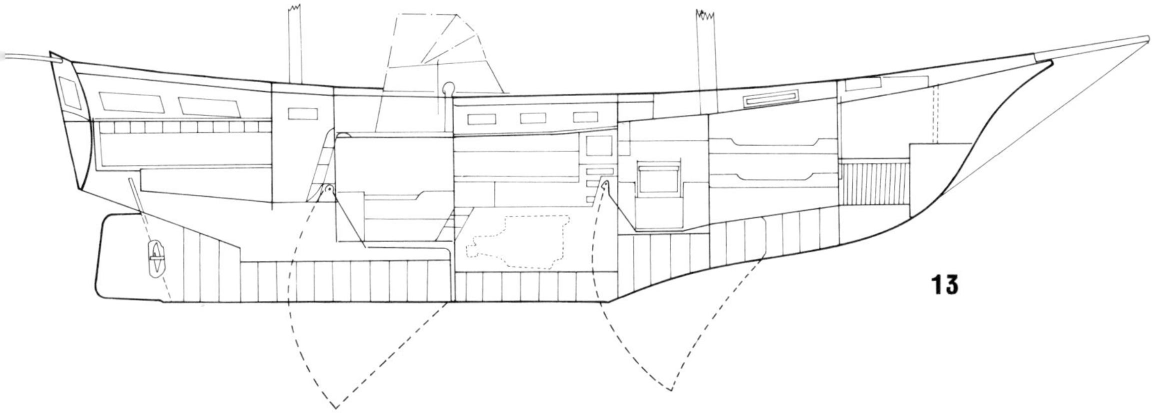

13

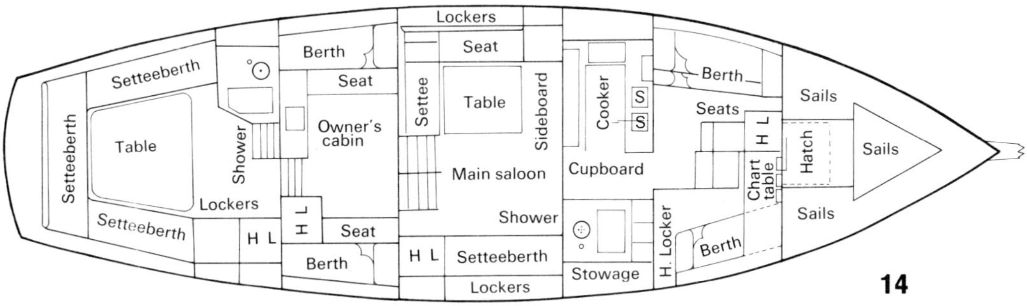
Lockers
Seat
Berth
Setteeberth
Seat
Settee
Table
Sideboard
Cooker
S
S
Berth
Sails
Seats
Setteeberth
Table
Owner's cabin
Shower
H L
Hatch
Sails
Main saloon
Cupboard
Chart table
Lockers
Shower
H. Locker
Sails
Setteeberth
H L
H L
Seat
Berth
Berth
H L
Setteeberth
Stowage
Lockers

14

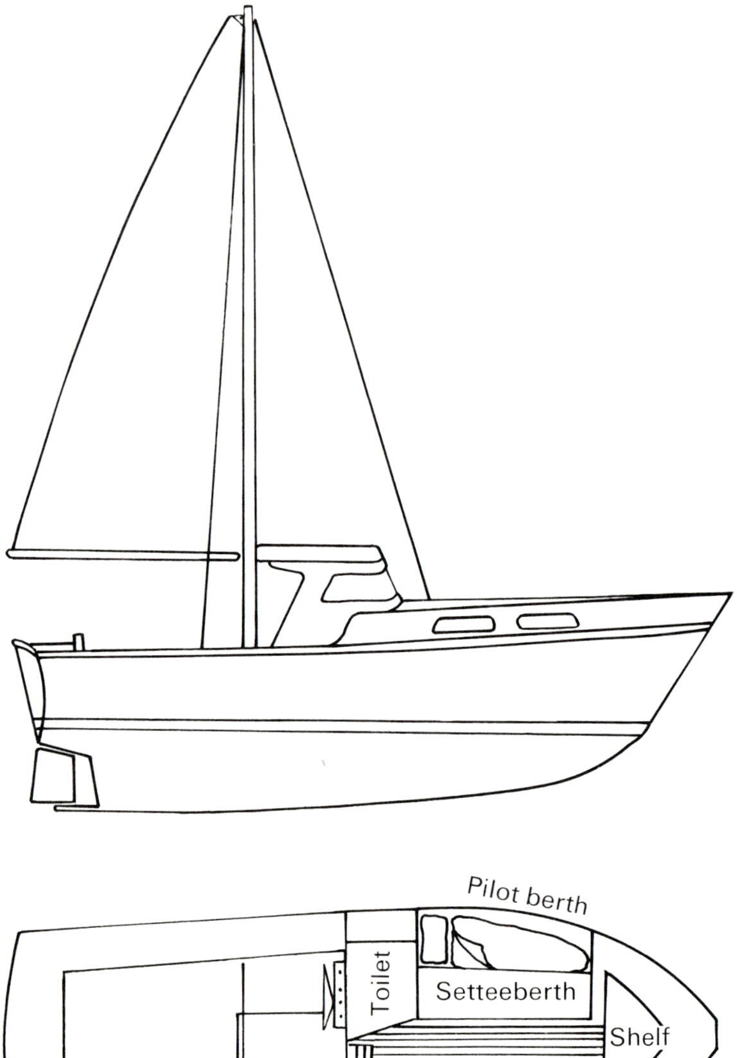

15

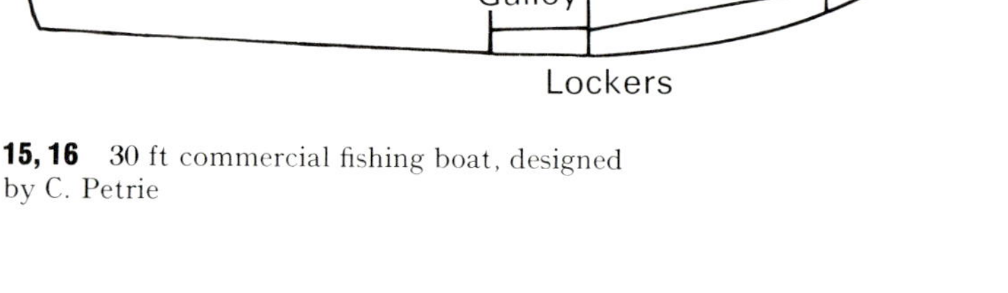

16

15, 16 30 ft commercial fishing boat, designed by C. Petrie

ferrocement 4

SELECTING A CONSTRUCTION TECHNIQUE

Pipe-frame technique
Male moulds
Cavity moulds

At this point the big decision has been taken and a suitable design chosen. What follows is the selection of the construction technique as this will to some extent determine the site requirements and the materials to be ordered. While the basic ferrocement concept remains essentially unchanged since it was defined by Nervi more than 25 years ago, incorporating as it does a steel rod and mesh matrix plastered with mortar, there have been some significant technological refinements.

In the past 10 years a number of different construction techniques have evolved, all as the result of experiments conducted by both professional and amateur builders in search of improved building methods, accelerated construction schedules and a more polished end product. The essential differences between the techniques lie mainly in the methods adopted to create the hull form and the rigid base to which is moulded the rod and mesh reinforcement; and which will later support the structure until such time as the implaced mortar is sufficiently cured. In most instances the framework is then removed, playing no further part in the hull structure.

At the present time the merits and demerits of each particular technique are the subject of some controversy among experienced ferrocement practitioners, who have been heard to condemn one or other method for various reasons. But despite these condemnations each technique which will be described below has had a reasonable record of success. Some of course appear to have more advantages than others, or may perhaps have more to offer either the commercial builder or the amateur, as the case may be. It is up to the individual to decide, after weighing up the pros and cons of each technique, which appears the most favourable in his eyes, and which will suit his requirements better.

Broadly speaking the different techniques can be grouped under three headings, each of which shall be discussed in turn. But it is the first technique on which I shall concentrate, as it is the most common framing-up method adopted by ferrocement boatbuilders, and is regarded as being more suitable for the homebuilder than any of the others, a fact which should become clear as we go along.

Pipe-frame technique

The pipe-frame technique can be described as the traditional method of building with ferrocement, for it was first developed by Prof. Nervi, and is undoubtedly the most common frame-up method adopted by ferrocement boatbuilders today. I can safely say that this method has been used to build most of the ferrocement craft thus far built in the western world, a figure which must now exceed 400.

With this method mild steel pipe, sometimes called steam or water pipe, is bent and shaped to fit the contour of each station frame. The piping varies in inside diameter between $\frac{1}{2}$ in. and 1 in. The same pipe, or sometimes of a larger diameter, is also used to outline the stem, keel loop and the stern section on a lofting floor. The skeleton is then erected and tacked and welded into place, extreme care being taken to ensure that all lines are fair and true. If a ferrocement deck is to be constructed, then piping may also be used across the beam of the vessel, shaped to the required camber. Cross-sectional braces will, in any case, be required to give the 'pilot' structure further strength and rigidity. In most cases this framework is built upright, but there is an alternative method, not generally advocated, of constructing the framework upside-down. This method has the complication of having to turn the hull right side up once the mortar has cured, which can be difficult if the vessel happens to be particularly large.

Once the hull skeleton has been properly aligned and braced the next stage in the construction is the placement of longitudinals or stringers. These are high tensile steel rods which are fastened horizontally onto the pipe frames at predetermined intervals. Attachment is made either by welding or with wire ties. Transverse or vertical rods, probably of a lighter gauge, follow, and are fastened to the stringers between the frames, as dictated by the design. Some builders consider them unnecessary, being of the opinion that the extra weight and greater hull thickness created by them is in no way compensated for by the increase in strength. This argument will be discussed at greater length later.

If the deck is to be of ferrocement, then the reinforcing rods can be set up in the same way as the hull. Alternatively rod starters can be left sticking out from the toerail, the deck to be laid up after the hull has been plastered and the mortar cured. The time has now come to lay up the mesh, which is probably the most tedious and time consuming job of all, necessitating the tying of thousands of wire ties, to lash together the reinforcement in a tight matrix. In most instances eight layers of mesh, which can be either welded mesh fabric which has been galvanised or the diamond pattern bird or chicken wire, are laid up, four on the inside and the other four on the outside, sandwiching the rods in between. The concentration of mesh layers can of course vary, but it must be remembered that there are defined limits for the steel content for ferrocement which should be adhered to. These limits have been discussed in previous chapters. In practise, they are achieved by selection of the right sizes of steel, the building up of a tight mesh structure, and maximum penetration of the mortar. Test panels can be made, before the hull structure is started, and weighed and tested for steel content and strength. At the time of placing the mesh, or shortly before, provision should be made for propellor shaft, rudder, chainplates and any other fittings which are to be incorporated with the hull structure. The hull is now ready for plastering, the mortar being applied from the inside and forced through the reinforcement. Finishing is carried out by plasterers on the outside.

As I mentioned above, the pipe-frame technique is the most widely used frame-up method adopted by ferrocement builders. In most cases it has provided reasonable results. One of the best known ferrocement yachts to be constructed in this manner is the 53 ft cutter *Awahnee*. A double-ended hull with canoe stern, *Awahnee* was designed by her owner, Californian vet Dr Bob Griffith, and built by him in New Zealand during the early 1960s. Since then she has completed a circumnavigation, going via Cape Horn, encountered numerous gales, and has come through with a clean bill of health. In *Awahnee's* construction $\frac{3}{4}$ in. diameter pipe was used for the frames and deck beams. They were bonded into the ferrocement structure and not removed after the hull was completed, as is sometimes the practice.

The principle objection some ferro boatbuilders have to the pipe-frame method is the problem of deflection inherent with the pipe-work. Water pipe is regarded as not being sufficiently strong or rigid, and when subjected to the weight of a large mass of wet mortar is likely to bend out of alignment unless sufficiently braced. Generally it is felt that while most

12 *Petralis*, a 65 ft ferrocement yacht launched into the Thames near Tower Bridge in October 1970. The vessel was entirely amateur built in just over a year by a team led by a young New Zealand architect John Aston. The design was originally meant for steel construction but was redesigned for ferrocement. The yacht has a beam of 15 ft and a 7 ft 6 in. draught. Her waterline length is 48 ft. $\frac{1}{2}$ in. Welded mesh and $\frac{1}{4}$ in. high tensile rod stringers were laid onto a pipe framework. Only six layers of mesh were needed to make up the desired steel concentration. Thickness of the hull was 1 in. As you can see the yacht is almost flush decked. Originally a ferrocement deck was planned, but this was waived in favour of plywood laid on teak laminated beams for greater weight saving above the waterline. *Petralis* was launched unballasted and her hull, which included internal wooden bulkheads, furniture and engine, touched 20 tons. A further 13 tons of ballast was dropped into her trough keel to bring her down to her marks.

amateur builders will support the hull structure just sufficiently to withstand the weight of mortar they often fail to realise the forces which are going to be exerted on the frames during the application of the mortar. Consequently there is deflection, as the pipes and then the mesh yield to the pressure applied to them.

Strong in his condemnation of the use of water pipe is the New Zealand boatbuilder Richard Hartley, who feels that they are 'responsible for the lumpy, bumpy hulls which gave ferrocement a bad name initially'. As an alternative to pipe, but which can be categorised under the same heading for the rest of the application is similar, Hartley devised the idea of prefabricating station frames using reinforced steel. To fabricate the frames two $\frac{5}{16}$ in. steel rods, in parallel, are bent to the shape of each frame, taken from a full sized template. Trusses are welded between them and, says Mr Hartley, they are 'strong enough for a man to step on during construction' or to take scaffolding during the plastering phase.

The same construction is adopted for the deck beams, with the whole system designed to remain within the hull after completion, forming an integral part of the structure. Mesh is wrapped round the protruding frames and plastered; they provide good fixing points for attaching bulkheads, engine beds and bunk framing and can save many hours of labour. Besides Hartley a number of other commercial builders are now adopting this framing technique. The mesh, stringers and transverse rods are applied in the same manner as for pipe bending.

As a further variation to this concept of providing station frames and moulding the reinforcement to them, some builders make use of wooden frames, made up in the same way as the framing for a conventional timber vessel. These are purely temporary frames and are removed once the mortar is sufficiently cured. In this application the four inner layers of mesh are first applied in the region of the frames, and roughly tacked to them. The stringers can then be placed round the framework and either stapled on or wired up to the mesh. The rest of the operation then follows logically as with the prefabricated and pipe-frame techniques. The advantages of this method are that wooden frames can be more accurately shaped to the contour of each station frame and any mistakes can be easily rectified. This is not often the case with either the water pipe or reinforcing rod which, once bent to a shape, is not easily rebent again without leaving distortions or ripples. Secondly, these wooden frames will not deflect or bend when pressure is exerted on them, and when properly braced internally with cross members will ensure against any gross deflections of the hull either when the reinforcement is being attached or when the hull is being plastered. One condemnation of the use of timber frames is that they will absorb water from the mortar during the critical curing period, which will leave weak streaks in the hull. I personally feel this is not a valid point, for the frames can be given a good soaking just after the hull has been plastered, and will be kept wet during the spraying of the hull. Absorption will be minimal, or the edges can be sealed.

Of course there are other problems of deflection than those related to the bending of the frames. Once the frames are properly braced, the chief concern is with the high tensile rods between them. A series of flats could be caused between the frames or ribs. This can be prevented by using more frames so that the longitudinals will follow a fair line round the hull.

A variation of this form of deflection could occur if, for example, the builder turns the rods round the bow or transom, continuing them along the other side. The usual practice is to cut the rods off at the stem and either tie or weld them there. What could happen is that when pressure is exerted on the turned rod on one side of the hull, say when the mortar is being applied, it could cause a hollow on that side and a resultant bulge on the other as the rod is forced out. This deflection could be missed until it is difficult to rectify. Where the reinforcement is erected in this manner care must be taken to coordinate activities when the mortar is applied, so that there may be a balanced pressure on both sides.

Another deflection danger is that of the mesh between the rods forming identations or long grooves between the stringers, which could occur when the mortar is applied. It is not always easy to stretch the mesh tightly so that it forms a rigid matrix which will not give as pressure is applied. These slight hollows can be built up carefully to give the hull a uniform appearance, but only by a skilled plasterer. There is the danger with this kind of filling, that if the unreinforced mortar layer is thicker than about $\frac{1}{8}$ in. it will be prone to flake off if subjected to a heavy knock, as when the boat may collide with a jetty or dock. The hull will flex under the impact and the thick layer of mortar will crack and flake, or at least tend to do so.

Partly in an attempt to counter some of these deflection problems, and so provide a boat which would not be likely to have that lumpy amateurish look associated with many ferrocement craft, the male mould or wooden plug was devised.

Male mould or cedar mould technique

Canadian ferrocement boatbuilder John Samson, who founded the company Samson Marine Design Enterprises Ltd, of Vancouver, is believed to be the first shipwright to adopt this system of 'pilot' framing, but merely as an alternative to the pipe-frame technique. A builder of wooden vessels before directing his activities to the construction and development of ferrocement craft, Samson had the idea of making use of some of his wood technology with this new medium, specifically by erecting a wooden plug on which to mould the steel rod and mesh. The purpose was to try and speed up production methods, reduce labour, and go some way toward solving the deflection problem, thus ending up with a much fairer hull.

Originally he used cedar for the construction of the plug, hence the name 'cedar mould'. Generally any reasonably soft wood will do equally well. The method involves creating an inverted hull form, constructed in much the same way as a conventional timber hull, but without the need for detail. First, station frames thick enough to provide a wide edge for nailing are shaped out of wood, taking the lines off a lofting floor. These are set up and nailed to a strongback with temporary braces or battens to hold them in place until the proper planking can be applied. The strongback could be constructed out of two long 4×12s set parallel to one another and held firm with trusses. These timbers should be slightly longer than the overall length of the hull.

Once the station frames have been correctly lined up and are standing vertical, the wood keel can be fastened on. Likewise the stern section or transom and the stem can be made up from the lofting floor and fastened to the frames and strongback. Now the job of putting on the planking, which can be 1×2 in. timber laths, can take place. Generally the form is to start from the toerail or sheer line, which will be near the floor, and work upwards toward the keel. A spacer should be left where the deck level is going to be so that reinforcing rod and mesh can be turned in as a starting structure for the proposed ferrocement deck or to construct a web for the wooden deck. Likewise provision must be made for starters, possibly near the frames, for the later construction of webs or flanges onto which bulkheads can be bolted. This can be done either by leaving spacers or gaps through which right-angled rod sections can be set, or holes can later be drilled into the planking and the rod starters inserted through them. After the planking has been completed the rough edges can be planed and the whole structure made as true and as smooth as possible. The completed form can then be covered with a plastic or polythene sheet, to close any gaps between the laths and to keep the mortar from bonding to the wood.

The different layers of reinforcement are now successively fixed to the mould. This can be done with a staple gun and pieces of wire. Four layers of mesh, either welded fabric or chicken wire, are attached first. The bottom ends of the mesh can be pushed in through the gap left for the deck starters and the rest spread out as smoothly and tightly as possible over the form. The mesh is stapled to the wood. Experience has shown though that it is not necessary to firmly staple the first four layers of mesh, but only the last welded layer. The upper layers of reinforcement will literally press the bottom layers into place.

The stringers and the transverse rods can now be attached, the former being placed first. Two sized high tensile rods are generally used, their diameters determined by the scantlings. The most common size for the stringers is $\frac{3}{8}$ in. diameter rod. These are laid on at predetermined centres, which may vary between 2 in. and 4 in. They are stapled into place, with their ends wrapped round the bow and transom. If possible they should be at least as long as the hull. Transverse rods can now be applied in much the same way as the stringers, overlapping at the keel and running right down to the decks. The rod starters for deck and bulkhead webbing can be placed at the same time. Finally the last four layers of mesh are stapled on. This must be well done to ensure that there are no bulges or loops of mesh protruding up, which might later cause a bump and unfairness in the hull. The reinforcement must be bound to the form as tightly as possible prior to the plastering, which is the next job.

One should understand at this juncture that there are variations of both reinforcement and the way in which it is applied to the mould. Another method makes use of a heavier gauge of mesh and reduces the number of layers to six instead of the usual eight. Two layers are applied first. These are followed by $\frac{3}{8}$ in. diameter high tensile rods, which are placed vertically, lapping the keel. Another two layers of mesh are applied and then the longitudinals are wrapped on and stapled down to the mould. These are of high tensile steel as well. It is felt that $\frac{3}{4}$ in. steel rod is the maximum thickness for the reinforcement as it is sometimes difficult to encase larger diameter rods with mortar.

As an alternative to rod one company has experimented with steel cable for the stringers or longitudinals. The cable is similar to that used for prestressing concrete and is made of seven strand wires giving a total diameter of $\frac{1}{4}$ in. The ultimate strength of this cable is given as 250,000 lbs per sq. in., as opposed to the $\frac{3}{8}$ in. high tensile steel generally used, which has a tensile strength of 75,000 lbs per sq. in. The cable is, as can be expected, slightly more flexible than the rod and is wound round the hull form in one continuous length at about 3 in. centres. It is a method which has shown some measure of success, providing an extremely strong and resilient hull. It should be noted that the cable is far more expensive than the reinforcing rod.

Once the longitudinals are placed the last two layers of mesh are fastened on and stapled firmly in place making as tight a matrix as possible. This is a vital part of the construction as it will to some extent determine the amount of mortar to be placed and consequently the weight of the structure: extra mortar does not necessarily mean extra strength.

One further point before plastering. Provision must be made for such hull fittings as the stern tube which takes the prop shaft, the rudder gland and brackets to hold the rudder shaft and its foot. In such areas holes can be drilled in the wooden mould, preferably of a larger diameter than necessary and the cavity reinforced with further mesh and rod. The brackets, if they are to be incorporated with the hull reinforcement, can be welded in. Other openings, like those for skin fittings, can be cut and wooden plugs inserted.

Plastering techniques for the male or cedar mould can follow three general patterns. In the first the normal mortar mix is placed in dollops on the formwork and forced through the mesh with gloved hands. A vibrator is often used in an attempt to get maximum penetration. A plasterer follows up with a trowel and moist sponge to give the hull a smooth and even finish.

Samson recommends a slightly different procedure which is divided into three stages. In the first a slightly wet, soft mix is applied with the idea of getting maximum penetration. A stiffer mix follows, and in both this and the previous application a vibrator is used. The final mortar application is to fill up those areas where there appears to be voids or a mortar deficiency. Work is concentrated on getting a good finish, so the mortar is kept fairly stiff, and is so worked over with a trowel to ensure that it only just provides a covering to the mesh.

A third method of application makes use of the Gunite process, sometimes known as shotcreting. With this technique the mortar is pressure-shot onto the hull fabrication, water being mixed in as the dry aggregate and cement mixture reach the nozzle. The force with which the mortar hits the reinforcement will ensure that there is good penetration. But the Gunite process is not recommended, for it usually necessitates too wet a mix, which results in weakening of the mortar and a large degree of shrinkage. Expert nozzle control is needed, and much heavier gauge mesh.

With the hull plastered the curing phase begins. During the initial days the mortar is continually sprayed with water so that the rate of hydration is maintained, allowing the hull to gain strength. Usually, following normal curing procedures and not, for example, making use of steam curing, about 28 days must elapse before any other work should be carried out on the hull. At this stage it is possible to sand the exterior to get an extra smooth finish, paint and prepare it with antifouling. This job is mostly left until the vessel is ready to go into the water.

Once sufficiently cured, the hull must be turned right side up. The wooden plug can now be ripped out. This must be done carefully to prevent micro-cracking or chipping of the mortar as the wood is pulled away from the staples. Exposed ends of these staples should be cut off short and ground away or they may cause rusting. Work on the interior can begin.

While adoption of the male mould has produced a few apparently flawless hulls it is not a technique which is recommended to the amateur. There are a number of pitfalls inherent with the process which make it less attractive than the alternative pipe-frame technique (under which category I also include the prefabricated frame). In the first place there is the extra cost and labour required to build the mould, a job which will need a certain amount of expertise at carpentry.

The most important problem associated with this technique is connected with penetration. There is no way of determining how effective are the efforts to achieve complete penetration of the reinforcement. The builder is working 'blind' and only after the wooden plug has been removed can the results be seen, at least to some extent. By then the hull is well on the way to being cured, and to detect and rectify the fault is not all that satisfactory. Voids are a feature which at all costs must be avoided as they will considerably weaken the hull.

Once the wooden mould has been removed the hull should be thoroughly tested for any voids. This can be done by tapping the hull lightly all over with a metal bar or ball peen hammer. Alternatively the metal bar can be run gently over the surface of the hull. Listen carefully for a change of tone. Should it become more resonant or deeper, it probably indicates a cavity. It is important to fill up the hole, for it is a weak spot in the hull which will not withstand much pressure or impact.

Once discovered, voids or cavities can be patched up relatively easily. Knock away the covering mortar to lay bare the hollows and brush away the loose bits of grit and dust. Large mortar chunks still adhering to the mesh can be left there. Straighten the reinforcement if it is bent out of shape. Providing the cavity is not moist a normal two-pot mix epoxy resin, given body with a quantity of neat cement or fine industrial sand, can be used as a filler. Combine the resin and the sand or cement to the consistency of a stiff paste and force it into the cavity using a gloved hand.

The filler can be worked in with your fingers and palms, but done in such a way to eliminate any likelihood of further voids being formed by trapped air. Another idea is to use a rich mortar mix, prepared in the conventional way. In this case the surface of the cavity should be coated with a good bonding compound. Epoxy resin can be used again. It will ensure a good bond between the new mortar and the old and at the same time act as a sealer and waterproofing layer.

The fact that the hull is built upside down presents another difficulty, especially for the home builder. Unless the hull is quite small, turning it will be quite a job if facilities are not available or are inadequate, which an amateur's are bound to be. I know of one or two builders who used this inverted mould method and later cursed their decision to do so.

Female or cavity mould

One of the reasons for the slow commercial development of ferrocement boatbuilding is that present techniques are labour intensive and more geared to building individual craft on a one-off basis than on a mass produced scale. And as labour costs tend to be high and are continuing to rise, prospective professional builders are dissuaded from embarking on a commercial venture. Or as the case has been, particularly in the US, they've begun a project with too little capital behind them, found out too late that they cannot keep pace in a highly competitive field, and have been forced to go into liquidation.

Of course some commercial ventures have been highly successful. Windboats Ltd of Wroxham have built up a thriving business using first the pipe-frame technique and later the prefabricated rod frame. Likewise Ferro-cement Ltd of Auckland, Samson Marine Design Enterprises Ltd, Vancouver and others are successfully keeping their heads above water. The Chinese are using the male mould technique to mass-produce hundreds of sampans.

Recently certain commercial builders have made considerable progress in the development of what appears to be a more efficient method of producing ferrocement marine structures with labour costs low enough to make commercial production more practical and profitable. As yet the technique is still in an early stage of development and cannot be said to have been proved superior to the other established techniques.

The new techniques makes use of a cavity mould, or female mould as it is commonly called. In principle it is similar to the female moulds used for the moulding of GRP hulls. One company which has developed such a

13 Fibersteel floating cavity mould, being inspected by Captain Irving M. Johnson, 1965. Photo: M. E. Iorns, Fibersteel Co.

technique and vouches for it is Fibersteel Co. of West Sacramento, California. Unfortunately they are unwilling, at this stage at any rate, to divulge too much concerning their process, except to their franchised builders, until patents have been received.

Briefly, their method involves the construction of a concrete female or cavity mould, formed originally from a prototype hull. The internal surface of the mould is made smooth. The reinforcement is fabricated on the inside of the mould in the proper sequence and mortar is then pneumatically applied. A releasing agent is initially applied to the mould so that, once the mortar has hardened, the hull can be floated out. Or alternatively, if fast production is required, the hull can be floated clear of its mould to be cured in another location·

According to the company there is no size limit to vessels constructed with the cavity mould: already there are designs for vessels well over 100 ft in length on the drawing board. Obviously the cost of producing a mould must vary depending on the size but it is estimated by Fibresteel that three completed hulls would pay for the mould. Thereafter the operation would become profitable. A further development of theirs is the production of a 'low-cost floating mould which is independent of any shipyard or launching facilities'. A photograph is published illustrating this mould. It makes possible the construction of ferrocement structures directly on the water.

This concept, if it is as successful as its developers expect, opens up new vistas for low cost construction of large concrete structures, other than boats, which may include breakwaters, wharfs, floating airports, marinas, etc.

Conclusion

As one can see, there are a number of methods which can be adopted for the construction of a ferrocement boat. For the amateur builder the cavity mould is obviously out. It is a method designed for the commercial builder whose aim is mass-production. The initial capital outlay and expertise required for the production of the mould makes it an uneconomic proposition for the home builder.

At first sight the cedar mould technique appears to offer a couple of advantages. It is a method which will reduce the problem of obtaining a fair hull by eliminating the danger of gross deflections caused by the partial collapse or sagging of the frame or reinforcement while the mortar is being applied, and until it cures. This feature will surely attract amateur builders who are desirous of having a fair hull without the indentations and imperfections which have, up to now, characterised many home-built ferrocement craft, or builders who are unsure of their ability to manipulate and adequately brace the pipe-frames or prefabricated rods, and prefer to take the safe way out.

The cedar mould is undoubtedly a much quicker technique than the more traditional framing methods used for ferrocement construction. For example, less time need be devoted to the laying up and fairing of the reinforcement. Take the case demonstrated by a group of novice builders who attended a dealership training course held by the Canadian firm Samson Marine Design Enterprises Ltd. As part of their course the seven men in the group had to build a 32 ft houseboat hull. This they completed to the point where it was ready for plastering in five days. Notwithstanding the fact that the wooden frames of the mould had already been constructed and fastened to the strongback, five days represents fast work. Obviously under normal working conditions production time would be somewhat longer, especially as there wouldn't be any experienced supervisor on tap to direct operations.

Weighing very heavily against this method of framing is the very real danger of voids being created within the ferrocement, or of inadequate penetration. While voids can to some extent be discovered by using a 'sounding bar' there is no real guarantee that all have been found out. The presence of cavities within the structure will seriously weaken the hull. Some builders have employed the Gunite process in an attempt to achieve good penetration, but this is a technique which is not recommended.

For the one-of-a-kind builder or the amateur it is my view that the pipe-frame technique, or variations of this method, is still the best answer. If the boat is to be of any size it is better built upright, thus alleviating the problem of turning the hull over once the mortar has cured. If water pipe is to be used, remember the dangers of deflection, not only from the weight of wet mortar but from the forces exerted on the reinforcement during the application of the mortar. It is worth spending those few extra days ensuring that the frame is well braced and the reinforcement as tight as possible before plastering begins.

ferrocement 5

SITE SELECTION

The first practical step toward the construction of your boat, no matter what the building technique, is the selection of a suitable site on which to build. Obviously, though, the building method will to some extent determine the site requirements. In GRP construction for example, a site must be chosen where careful control can be made of temperature gradients and humidity levels.

In making your selection study carefully the economics and logistics of each site being considered. Determine what facilities you will need and what are available. An ideal site would be on or alongside water somewhere so that when the time comes for launching you are not going to be faced with the problem and expense of transporting the hull any great distance. Unfortunately such sites are often hard to come by or rentals can be extremely high, enough to dissuade most amateurs working on a tight budget.

On the other hand don't plug for low cost land alone which could be well away from water. The expense in transporting a large hull to water can be extremely heavy. I know of one builder who fell into this trap and ended up having to pay transport costs which almost equalled the total price of materials he used in the construction of the boat's hull. In the end most amateur builders opt for their own back yards if space allows. That odd moment off can productively be spent on some aspect of the vessel's construction instead of travelling.

Unlike most other construction methods, it is not absolutely necessary for ferrocement hulls to be built under cover. No serious problems arise from exposure to the elements, other than rusting of the basket and possible washing out of cement from a newly plastered

14 The first practical step toward the construction of the boat is the selection of a site on which to build. Ideally it would be alongside water somewhere so that launching will present few transport problems or costs. With ferrocement boatbuilding it is not absolutely necessary for the hull to be constructed under cover, although some sort of protection will make the work more pleasant and allow it to continue in all conditions. An ideal site, alongside water, at Brentford in West London where Chris Cole has been building a 63 ft ferrocement yacht. A prefabricated wooden shelter covered with plastic sheeting was erected to provide cover. Photo: Chris Cole.

hull. Nevertheless, if a large shed, boathouse, or even a deserted warehouse can be acquired so much the better. If the aim is to build in the hope of obtaining a Lloyd's classification you will inevitably require a shed of some sort where conditions can be properly regulated. You will, in any case, need a completely dry shelter for cement, and means of protecting your aggregate from contamination.

Working conditions will be all the more comfortable if you have a roof over your head. Productivity will not be impeded by poor weather conditions. But make certain that the building is going to be sufficiently large to accommodate the completed hull and still provide adequate space all round it to take the scaffolding from which work will be carried out on the hull, especially during the plastering phase. A support structure is going to be needed to hold up the frame of the vessel during its construction. If the shed has beams, rafters and uprights these can form part of this structure. Allow a good clearance above the deck level of the boat. About six feet would be sufficient. Remember that the hull is going to be lifted onto a trailer when the time comes to move. The importance of a shed or similar protection from the elements is governed to some extent by prevailing weather conditions. In an area where you are likely to have continuing rain storms, snow and cold weather, a shed of some sort will have tremendous value. In warmer climates it won't be so necessary.

As you are going to erect a framework anyway to support the hull, one might just as well go one step further and so design the structure to allow for some protection overhead. A covering will be needed later during the construction to protect the hull while the mortar is still green and curing. Common practice is to erect a simple wooden frame over the hull onto which is draped polythene sheeting or canvas.

A firm level base is preferable on your construction site. A concrete or wooden floor would be ideal. If you have to make do with a soil base it would be a good idea to compact it as well as you can in an effort to reduce the likelihood of subsidence and movement. Neat cement or lime liberally spread over the area and lightly sprayed with water often helps to consolidate the ground. Avoid building at the bottom of a decline or in an area of low ground unless your drainage system is good. In a climate such as we have here in Britain it would not take long for your construction site to be turned into a quagmire if the necessary precautions are not taken to get rid of the accumulated rainwater.

If you don't happen to be building under cover some provision must be made for the storage of equipment, and materials until they are used. Cement, for example, will quickly deteriorate and become unusable if it is exposed to dampness. It is usual to have the cement delivered only shortly before plastering is likely to commence. Ensure that a power point is handy on the site. Electricity will be needed to operate the various power tools, grinders, saws and drills. If an arc welder is used instead of oxy-acetylene and rod, that is going to need power as well. Sufficient lighting will have to be wired up, for you don't want to limit your production to daylight hours.

A good and regular supply of water is essential. A lot of water is going to be required for the mixing of the mortar, and a great deal more is going to be needed to keep the hull moist after plastering to regulate the curing process. Particularly where the mixing of the mortar is concerned, not all water is suitable. No substances must be contained in the mixing water that are likely to have an adverse effect on the setting rate or quality of the mortar.

Any silts or organic impurities which may be dissolved in the water could be harmful. Water which contains such materials as clay, loams or acids is unsuitable. Generally the water drawn from regular sources like town supplies is quite all right. Where one has to make do with an alternative source, i.e. from wells or rivers, such water should be regarded with suspicion until a few samples have been drawn and tested. There are few rivers anywhere which are not polluted! Other than laboratory tests, one practical method of roughly determining the suitability of a particular water supply is to make a comparison of a few plaster samples, where some are made with distilled water and others with the 'suspect' liquid. If there is no apparent difference in quality between them then one could reasonably assume that the water is safe to use. Water which is brackish or seawater should not be used. It has a tendency to make concrete hydroscopic and will also increase the chance of the reinforcement becoming rusty. As a last resort salt water could be employed during the later curing stages to spray the hull, if one had to choose between no water or salt water.

ferrocement 6
CONSTRUCTION

Pipe
Rod
Mesh

Assuming that a suitable site has been found, only one more preliminary job requires doing before the real business of building can begin. This, logically, is the acquisition of all the necessary items of equipment and materials to be used, in construction. The primary materials which are needed at this early stage are the pipe, steel rod and mesh which together make up the reinforcement of the hull. The cement and aggregate, the other primary components used in the hull structure, need only be ordered when construction nears the plastering stage. Only fresh cement must be used for the mortar, the reason being that cement deteriorates over a period of time, particularly in a damp atmosphere. However, especially with regard to the aggregate, one can always explore possible sources of supply. It would save a lot of trouble later.

In a previous chapter a general description has been given of some of the more common construction techniques employed by ferrocement boat builders. It is not my intention to now give a 'blow-by-blow' account of each method, but rather dwell on those which are the most suitable for the home-builder, who will obviously have limited resources and experience of this new medium. These techniques revolve essentially round the upright open frame system, which includes the use of pipe-frames, prefabricated frames, or wooden station frames for the hull form and the support of the rod and mesh membrane.

Of these three techniques, the prefabricated rod frame method has more to recommend it than the other two. It is a frame-up technique which is much preferred by most experienced ferrocement technicians, and should advisedly be your first choice. If properly done the method will largely eliminate the dangers of deflection inherent in the pipe-frame method, and it will provide webs within the hull to which bulkheads, engine beds, bunk frames and other internal joinery can be fixed.

The pipe-frame technique would be your second choice. It has been successfully used to produce a large number of concrete craft. However, if pipe-frames are to be used the dangers of deflection must be remembered and adequate steps must be taken to ensure that the hull is well braced. Some builders condemn the technique for this reason. They also point out that some provision must be made within the hull, in the form of webs or flanges, for the fixing of bulkheads. As these will have to be fabricated in addition to the pipe frames, one might just as well use prefabricated frames. The only real advantage I can see in the use of pipe instead of rod frames is that they can be more easily and quickly shaped to the hull form. For the home builder who is a novice where welding is concerned the fabrication of the rod frames might present a few difficulties. But no doubt these can be overcome with a bit of practice.

The third technique makes use of temporary wooden frames, constructed in much the same way as the ribs for a wooden vessel, and removed once the hull has cured sufficiently. The frames are relatively easy to build and will provide a rigid base on which to fabricate your rod and mesh armature. But like the pipe-frame method the wooden frames do not allow for the later attachment of bulkheads, etc. Such fixing points will have to be fabricated later. Then, too, the frames are going to be removed once the hull is completed. The prefabricated frames are designed to remain once the hull is completed. The prefabricated frames are designed to remain within the hull; the pipe-frames can either be enmeshed within the structure or removed. The general view is that they are best left in the hull, to add to the strength and rigidity of the structure.

Now to return to the selection of the steel needed to make up the reinforcement. Other secondary materials which will be needed during construction will be discussed as they are required.

Pipe

The material used for the construction of the pipe-frames is variously described as black water pipe, steam pipe or galvanised pipe. This is a very general label: naturally there are pipes of varying quality which will meet this description. To ensure that you get the best quality select pipe which complies with a suitable specification. It should be readily obtainable from most dealers specialising in plumbing supplies. For craft up to about 70 ft length over all, pipe of between $\frac{1}{2}$ in. and 1 in. internal diameter has been used, with the most common size being $\frac{1}{2}$ in. internal diameter. Occasionally builders have used much larger diameter pipes, but these have been limited to such areas as the stem, keel backbone and sternpost. Pipe lengths vary but the most convenient dimension is 20 ft.

The quantity of pipe needed for the job will obviously alter from project to project. It is dependent on the size of vessel, her design and a number of other variables. Remember that this framing method is prone to deflection so it is going to need strengthening. Pipe is often used for cross braces, for the formation of deck beams and for verticals. Consider every possible area where you might need it. You might just as well buy the lot at once for you could obtain a substantial discount. Every saving that can be made on a costly project like building a boat is a considerable help.

Make allowances for a certain amount of wastage. If possible whole pipe lengths should be used for the construction of the frames (each half breadth). It is better to crop off a section than make up the required length by welding pieces together. Welding will produce a potentially weak area, which could give way when heavy loads are exerted on the hull. The consequences would be serious. And it will be difficult to weld a proper joint without creating distortions in the pipe.

Steel rod

Two basic types of reinforcing rod are used in the construction of a ferrocement hull. They are mild steel rod and deformed or high tensile steel bar, the latter being a refinement of the mild steel having improved strength characteristics. The mild steel is a hot-rolled bar having a low carbon content and can be readily bent and welded. Because of this characteristic it is often adopted for prefabricated frames. The rod selected should comply with BS 4449, in which the minimum yield stress is in the region of 30,000 lbs per sq. in. It is, of course, available in different dimensions. For vessels up to at least 70 ft l.o.a. $\frac{5}{16}$ in. diameter rod is a practical size. The rod is manufactured in stock lengths up to 40 ft. Manufacturers are prepared to supply to order within this limit. Mild steel rod should not be used for the stringers or transverse rods in a ferrocement hull. Wherever possible high tensile rod should be used for these.

High tensile steel rod, sometimes described as deformed bar, starts life as mild steel. The HT rod can be formed in two ways. In the first it is further hot-rolled, during which process its carbon content increases, accompanied by an increase in tensile strength. In the second process mild steel rods are cold drawn or stretched or twisted. This effectively raises the tensile strength and yield strength of the material. The rod also becomes more springy and hard. However, strength gained is always accompanied at the expense of ductility, and the HT rod becomes less easy to manipulate and bend than mild steel, and more prone to fracture during bending. A further limitation is that a loss of strength will occur if the rod is heated. If the rods are to be welded, therefore, it should not be done indiscriminately: indeed it should be done as little as possible, and with the minimum amount and duration of heating.

Strengthwise there is nothing to differentiate between the two types of HT rod. Provided they meet their relevant specifications, BS 4449 for the hot-rolled HT rod and BS 4461 for the cold drawn rod, the minimum yield strength exhibited should be in the region of 60,000 lbs per sq. in. The cold drawn deformed bar can often be identified by diagonal ribs running down its length. In the formation of these rods square-sectioned mild steel rods are twisted and stretched, a process which creates the ribs. However, this is a dubious way of distinguishing between the two types of rebar. Some manufacturers also rib the hot-rolled rod, to provide better adhesion between the mortar and the imbedded reinforcement. The adhesion of mortar and steel is merely a measure of the friction which is created when pressure is introduced by the shrinkage of the concrete round the steel. The ribs offer improved holding.

High tensile steel rod is the most suitable material for the horizontals and transverse members. In the first place it has greater strength than mild steel. It is more springy and harder, and will better take up the curvature of the hull and hold its form, thus ensuring a fairer hull in the end. It will not sag between stations or kink round them when stretched taut as mild steel has a tendency to do. HT rod is sometimes used for the construction of prefab frames, but I cannot see the real sense in this. It is more difficult to mould into the complex form needed, and the excessive welding will probably reduce its strength to that of mild steel.

Naturally the selection of the rod dimensions must be governed by the scantlings of the structure. Unfortunately, at the present time there is no authoritive code of practice which can give a lead in this matter. Builders, and designers, new to this medium either have to resort to representative test panels to determine strength properties, or rely on the advice and experience of those who have practical knowledge. Like mild steel rod, deformed bar is offered in stock lengths, but manufacturers are prepared to crop rod to the buyer's requirements. Forty feet is the maximum length normally supplied. In selecting HT rod ensure that the material is straight without kinks and distortions, for these blemishes will be almost impossible to remove and they will invariably create difficulties with the fairing of the hull. Diameter of the rods plays a significant part in determining the thickness of a hull, and hence the weight of the structure. The thinner the hull the better from a weight point of view. However this reduction in cross section is accompanied by a loss of rigidity. You will find that, using present day materials, hull thickness of vessels between 35 ft and 50 ft will be in the region of $\frac{3}{4}$ in. Above this length it will probably increase to an inch or more. For ferrocement hulls between 30 ft and 70 ft spacer rod dimensions remain fairly constant, with diameters limited to between $\frac{1}{8}$ in. and $\frac{1}{2}$ in. In practice the most common diameter rod used for the horizontals is $\frac{1}{4}$ in. Transverse rods are usually of a much smaller diameter, probably about $\frac{3}{16}$ in. or $\frac{1}{8}$ in. The weight per foot of the rod can be obtained from the suppliers. For $\frac{1}{4}$ in. rebar it is generally in the region of 0.167 lbs per ft.

Mesh

The suitability of different types of mesh for use in ferrocement hulls has been one of the most controversial issues associated with this construction technique. As we know, the function of the mesh and rod is, in the first instance, to act as a lath providing the hull form and to support the mortar in its 'green' state. Then its function is to absorb the tensile stresses on the hull which the mortar, on its own would not be able to withstand. This follows from the fundamental fact, from which the ferrocement concept evolved, that concrete will withstand large strains in the neighbourhood of any reinforcement. The ability of a reinforced structure to withstand specific strains is dependent on the distribution and subdivision of the reinforcement through the mass of concrete.

Basically the strains on the hull are created from two sources: from without the structure and from forces within it. A hull is subject to a great deal of pounding, twisting and bending during its lifetime. This movement creates tensile stresses which mortar on its own is unable to withstand. Cracks or fractures will appear unless sufficient steel reinforcement is introduced to absorb these stresses. The degree to which fracturing of the hull is reduced is dependent on the concentration and dimensions of the imbedded reinforcement.

From within the structure stresses are set up by the mortar as it dries and cures. It is a fact that a mortar rich in cement, as is that used for ferrocement, is apt to have a high degree of shrinkage and creep. As this shrinkage occurs, so strong tensile stresses are set up. The mortar will crack if unreinforced. However, the inclusion of a uniformly distributed mesh fabric will absorb these stresses, limiting the degree of cracking. Clearly the more finely distributed the steel the smaller and fewer will be the fractures or cracks.

In selecting a mesh, therefore, all these facets must be borne in mind. Unfortunately there is no mesh on the market which will comply with each particular facet equally well, and one has to compromise. As a lath the need is for a mesh which is rigid from which to 'hang' the wet mortar, and which will not easily be bent out of alignment by the weight of the mortar or the force of its application. But then it must also be sufficiently pliable so that it can be manipulated into forming the compound curves of the hull. Then again it would be desirable to have a very finely woven and closeknit mesh which will reduce the micro-fracturing of the mortar to an absolute minimum. Yet the spacing must not be so close as to prevent proper penetration by the mortar.

From practical experience and experimentation Nervi, Collen, Byrne and others have determined that in order to produce a ferrocement structure with good mechanical properties the density of the material should be about 151 lbs per cu. ft. Of this the quantity of mesh should not be less than 24 to 36 lbs per cu. ft, or about 15 per cent. This, in effect, would mean the provision of layers of mesh giving a total weight of 2 to 3 lbs per sq. ft of surface area for an inch thickness of the ferrocement shell. While it will not greatly effect the structure if the weight of mesh is more than this, it is important that the level does not drop below it. Strength will drop quickly.

In an attempt to meet these requirements various types of mesh have been used for the construction of ferrocement hulls, and continuous development work is being carried out to improve the design of mesh fabrications. A few commercial builders use designs specially fabricated for them. Prof. Nervi, for example, used a standard type of mesh available at that time which was often used for the construction of ceilings. This was a mild steel wire composition, the wire being 0.02 in. to 0.06 in. diameter. The mesh was a square fabrication, the holes being 0.4 in. Weight of the mesh was in the region of 0.14 to 0.35 lbs per sq. ft. The New Zealand built yacht *Awahanee,* the first concrete yacht to circumnavigate the globe, also had a square welded mesh in her structure.

Present practice makes use of two principle types of mesh. The first and probably most commonly used type is often described as bird netting or chicken wire. The main reason for it being the popular brand of mesh is that it is relatively cheap; at least cheaper than the other fabrication. It is also more readily available. Chicken wire mesh is fabricated from cold drawn wire which is woven into a hexagonal or diamond pattern fabric. It is also galvanised, either before or after fabrication. Chicken wire which hasn't been galvanised is also available, but is not recommended.

Mesh is fabricated in different sizes and gauges. Builders using this type of mesh generally recommend 18 to 22 gauge with mesh openings varying between ½ in. and 1 in. Anything much smaller than ½ in. makes it difficult to get complete penetration of the steel matrix. Mesh much larger than 1 in. does not give a sufficient concentration of steel. The prevalence of fractures or cracks increases. Netting can usually be obtained in rolls 3 ft by 150 ft.

The second type of mesh used by many builders is welded mesh. It is a bit more expensive than bird netting, but is also quite a bit stronger, and able to withstand bigger tensile stresses. This material may be galvanised It is a much stiffer mesh and some builders prefer it for they consider that it will mould more easily to the curves of the hull, producing much fairer lines. Hollows and ridges are less likely to be created during the application of mortar than with the bird netting. Like the bird netting, welded mesh is available in convenient 3 ft rolls 150 ft in length. The most common dimensions for this mesh are ½ in. squares the wire being either 18 or 19 gauge. BS 4483, sometimes quoted, applies to reinforced concrete for buildings, for which fire regulations require different steel treatment.

One point concerning the mesh which has been raised among the so-called ferrocement pundits has been the pros and cons of using fabrications which have been galvanised. Prof. Nervi and the other early ferrocement technicians used ungalvanised steel. Their structures have been successful. What has concerned some builders here is that where there is a mixture of plain and galvanised steel, such as in the rod and mesh membrane, there is a possibility of a galvanic couple being formed between the two metals. But as yet there has been no evidence to indicate that such a reaction has occurred or that the result has had any detrimental effect on the structure.

Also, any galvanised metal will be attacked by caustic solutions such as calcium hydroxide. This chemical is present in mortar, so galvanised mesh will be attacked while imbedded in the fresh mortar; the moisture and the hydroxide will react with the zinc galvanising to form calcium zincate. The mesh will begin to corrode. But it appears to be a far from serious effect. Evidence indicates that only the surface layer of zinc is eaten away, for the film of zincate which is formed provides a protective layer for the underlying metal. It is also considered that the resulting keying of the surface could be advantageous in bonding the steel to the concrete.

There is a third alternative to the bird netting or welded mesh which ferrocement technicians sometimes use. This is known as expanded metal or metal plasterer's lath. It is formed by cutting a thin gauge steel sheet to give diamond shaped openings when the metal is stretched out. The manufacturing process is much less labour intensive than the method used for manufacturing bird netting or welded mesh, which results in a much cheaper material. Investigations into the possibility of using expanded metal as an alternative to mesh fabrications in ferrocement were carried out by J. G. Byrne and W. Wright at Trinity College, Dublin. The experiments were initially carried out to determine the suitability of the material for a roof canopy. Prof. D. G. Collen likewise carried out a series of experiments with expanded metal at the same university.

In Collen's work sample slabs were constructed of mortar and expanded metal and a series of bending tests carried out, the results being compared with those obtained for similar tests on the woven mesh. Collen observed that, weight for weight, the expanded metal was not as strong as the woven mesh. But for a cost to strength ratio the expanded metal had an advantage. Byrne and Wright came to the same conclusions. In one tensile test with ½ in. expanded metal the mortar cracked badly. This was put down to 'a scissor action of the diamond mesh'. They believed that this indicated a limit to the size and weight of expanded metal which can be used. Later tests with a lighter material with ¾ in. openings provided more satisfactory results.

Expanded metal will not shape as easily to the compound curves of a hull. Another practical disadvantage of this material is that it also increases the difficulties of obtaining good mortar penetration. Against this is the fact that being heavier than the other fabrics few layers are needed to make up the desired weight of steel within the structure. In general it is not recommended to the amateur builder who has never worked with mesh fabrications before, although it requires less labour.

British Reinforced Concrete Engineering Co., Stafford, Staffordshire, have done considerable research and development on thin-section ferrocement for hulls. They will willingly provide technical information on steel reinforcement and suitable mortar mixes for ferro craft.

ferrocement 7

LOFTING

Pipe-frame fabrication
Prefabricated rod frames
Wood frames
Setting up

The process of reproducing to full size the architect's scale drawings and lines of a boat design, and the shaping of the framing material to create the hull form, is known as lofting. No matter what construction technique may be adopted this job must be carried out first. Obviously it is a vitally important phase of the construction and extreme care must be taken to ensure that all lines are accurately reproduced. Any inaccuracies at this stage, if not rectified straight away, will be carried right through to the completion of the vessel and could very well determine the success or failure of the project.

Many laymen, perhaps understandably, imagine that lofting the lines of a boat must be an intricate and complicated job. However, it is more straightforward than one would think, although it does require a great deal of care and attention to detail. Any amateur tackling the job for the first time and unsure of his ability to carry it out properly would be well advised to read a few of the many books on the subject, and enlist the services of a naval architect or experienced shipwright who can provide a guiding hand or advice when it is needed. But for a start here are a few guidelines.

Every set of plans includes a lines drawing, construction plan, general arrangement plan, deck plan, sail plan and so on. For the development of the hull form all that is needed of these plans is the lines drawing and with it the table of offsets. This latter is merely a collection of measurements, or coordinates, which indicate points on a grid. These positions marked and connected will reflect the transverse station lines of the hull and the hull profile. Further figures given with the table will indicate the position of each station along a fixed horizontal line known as the waterline. The hull can be seen to take shape when the constructed station frames are erected in correct order and connected by horizontals or stringers.

If by chance you haven't been provided with a table of offsets, which is unlikely, you will naturally have to scale up from the drawings, working out the coordinates for yourself. Once again accuracy is essential and the drawings, being on a relatively small scale, must be closely followed. Pencil lines on the original can often mean as much as $\frac{1}{2}$ in. when scaled up to the desired size. Faced with this job, some builders will loft the hull to half size, producing a model, from which all measurements can be checked. The form can be examined for fairness. Check and recheck all your figures again and again. It is well worth the effort.

You are going to reproduce the scale drawings to full size, so obviously you will be needing a clear area in which to carry out the job. A smooth level surface is important. The concrete floor of a two-car garage, providing that it is clean and clear of oil spillage, will do the trick, and will enable craft at least 50 ft in size to be lofted out. A better idea is to use a few sheets of scrap plywood. About five or six sheets of 4×8 ft ply laid out to give good floor coverage, and firmly held together on a wooden frame, will be sufficient for yachts up to about 60 ft. The advantage of this structure is that it can easily be stacked up against a wall out of the way when lofting is not in progress.

If at all possible attempt to carry out the lofting under cover. Your work schedule won't be dictated by the weather, and the danger of having rain warp your lofting floor, distorting your lines, will largely be eliminated. You will also need good lighting so that you can maintain a high degree of accuracy. Poor conditions will surely effect the quality of your work. It is a good idea to paint the surface of your lofting floor in a light colour, white or grey, for ease of marking and to enable the lines to be better seen.

In order to plot your coordinates on the lofting floor a grid will be necessary. The dimensions of this grid, composed of horizontal and vertical lines, will be provided with the table of offsets. To prepare it you first set down three basic lines. These are the centreline, which is a vertical plane running through the centre of the boat from bow to stern; the baseline or datum, which is an arbitrary position on a horizontal plane usually above the sheer of the deck; and lastly the waterline, also a position on a horizontal plane which reflects the designed draught of the vessel. With these drawn in the rest of the horizontal and vertical lines can be drawn up. The horizontals are often described as waterlines and are positioned at regular given intervals down the centreline, measured from the baseline. The verticals indicate the buttock lines, and they too are positioned at regular intervals, this time along the baseline.

You are now ready to plot your coordinates or offsets, which in architectural terms are described as being the 'height to baseline' distance and the 'half-breadth' distance. It is usual that the offsets indicate the outer surface of the hull. As the frames take the inner surface of the hull, the shell thickness should be allowed for when drawing out your lines. In a vessel between 30 ft and about 50 ft l.o.a. hull thickness would be close on $\frac{3}{4}$ in.; from 50 ft to about 70 ft work on at least an inch. Measurements given in the table are usually in feet, inches and eighths of an inch. Sometimes, when a sixteenth of an inch may be added this will be indicated by a plus sign.

When starting to plot the offsets it is a good idea to begin with the largest station, for it will enable you to determine whether the loft floor is going to be large enough. With the coordinates for the first station plotted, go back and check them once more, to ensure that you have been accurate in each case. Mark the intersection clearly so that you cannot mistake its position. Following this you can now plot the coordinates for all the other stations.

With your 'half-breadths' and 'height to baseline' distances measured off, one more set of coordinates have to be plotted. These are the diagonals (also provided with the table of offsets) which indicate the sheer lines. Mark the positions clearly. Once all points have been marked, knock a nail into each position. They should be vertical and firmly imbedded in the loft floor so that they won't work loose. Now comes the job of connecting each of the points to outline the shape of the station frames. For this you are going to need a couple of thin battens which can take the curves. Straight-grained strips of wood about 4–5 ft in length would be suitable.

Place the batten on its edge and press it against the nails indicating the line of the station. Choose the concave curve as it will allow the batten to take up the necessary shape. This operation must be carried out carefully and the line adjusted to ensure a smooth curve. It might be an idea to have an assistant along to hold the batten in place while the station line is drawn, joining the coordinates. A soft-leaded pencil, with the point given a flat edge, can be used; or a felt-tipped pen, if the lofting floor is under cover. Any moisture might cause the ink to run and smear.

Besides the station frames the stem, keel and stern (which includes the horn piece and sternpost) have to be lofted as well. Often the practice is to set down the lines for these sections on the loft floor at the same time as the stations, superimposing the profile over the transverse lines. But with the danger of getting mixed up and confused with the many different lines criss-crossing the loft floor, it is advisable to first finish the shaping of the materials used for the station frames before going on to this job.

If your working space is somewhat limited, as it probably will be, then you will most definitely leave the lofting of the profile until after you have completed the shaping of the transverse sections. Or, of course, you could well have started with the profile and ended with the stations. You will also have to loft out the keel if it is a developed section. If the deck beams are to be integrated with the station frames, as with prefabricated frame, then you will also have to shape these members as well before the whole affair can be set up. The radius for the deck and the cabin top will also be included with the table of offsets.

Pipe-frame fabrication

After completing the lofting out of the transverse lines of the boat's hull, and after they have been thoroughly checked for accuracy, you can get down to the business of forming the frames. Start either from the bow (station 1) or the stern (the last station or transom frame) and work systematically through to the other end. And as each pair of station frames is completed mark them and stack them in order, to prevent the possibility of mixing them up.

Assuming that you are using the material recommended for the pipe-frames ($\frac{1}{2}$ in. to $\frac{3}{4}$ in. galvanised steam pipe) various methods can be employed to shape the frames. The usual type of pipe bender used by plumbers is probably the most common tool used and can either be borrowed or hired for the job. One builder uses a hydraulic portable bender (see photograph). If these tools are not available, don't panic. It is simple enough to build a pipe bender of sorts. One idea is to obtain a pipe T-joint, with the internal diameter at least $\frac{1}{2}$ in. wider than the frame pipe. Screw another length of pipe to the base of the T-joint, making it long enough to give you good leverage when it comes to shaping the frame. Two of these benders used in conjunction with each other will enable you to work more easily and efficiently.

Another method makes use of a length of timber, 5–6 ft long with width and thickness sufficiently large enough to provide strength. Bore a hole through one end, the diameter being somewhat larger than that of the pipe. This tool can be used in the same way as the T-joint bender. A technique we have used successfully from time to time, which is similar in principle to the commercial pipe bender, incorporates a $\frac{3}{4}$ in. steel plate to which is welded short lengths of rounded steel bar. The steel bars are welded upright to the plate about 3 in. apart. This must then be placed on a firm base which will not shift when pressure is exerted to bend the pipe. A vice attached to a workbench will do the trick. If you are using the hydraulic bender or any other portable apparatus, the pipe bending can probably be carried out directly on the loft floor, which is convenient for making frequent checks on the station outline. Take care when doing this not to damage the loft floor or erase any of the other lines. A few thin strips of wood on which the pipe can rest would prevent this possibility. If you are not working directly over the line, attempt to get your pipe bender as near as possible to it, and perhaps close to the ground. You will frequently have to drop the pipe on the loft to check that you are closely following the required curve.

The length of the pipe does not really matter at this stage. Obviously if it is going to be far too long for the particular station frame you're working on, and its length will hinder the shaping operation, then it is best to crop it. A hacksaw will do this though a proper pipe cutter will make the job easier. Generally pipe lengths are available in 20 ft sections. In determining the length you are going to need for a particular station frame bear in mind the design. Make allowance at the base for the height of the keel frame, horn piece, stem or whatever. Decide whether you are going to extend the pipe above the deck line to provide framing for a high toerail or bulwark. If the intention is to have a narrow ferrocement deck, or a web on which to attach a plywood deck later, some builders suggest bending the pipe frames at deck level to take up this web. The idea is not recommended, for the angle will be fairly acute and can present problems. In such a case it would be more practical to weld on separate beam frames. It is good practice always to have the pipe length a bit longer than necessary. It is always easy to crop off a bit, but having to add on a section is likely to present problems.

In shaping your pipe-frame do not start from the middle but from one of the ends, preferably at the keel, and work along to the other. If you start somewhere in the middle you are going to find it extremely difficult to make minor corrections without putting the ends out of line. Before starting, to make the shaping a bit easier, knock a few nails into your station line bringing it into relief. Remember to bend the pipe along the inside of the outline, closest to the centreline. The outer edge of the pipe-frame will take up the inner surface of the hull shell. As the shape and fairness of the hull is going to depend on the accuracy with which you

15 Various pipe-bending aids have been used. A conventional plumber's pipe-bender or electrician's hickey are perhaps the most common. Here a hydraulic pipe-bender is being used.

16 The stem pipe is carefully bent along its lofting line, drawn out on the loft floor. Wooden blocks nailed to the floor on either side of the line proved useful for holding the pipe in place.

17 With the frames lofted and bent to shape they can be erected on the site to form the framework. Here the keel loop and stem are in place on a firm base of railway sleepers. Note the layers of mesh for the keel already set out, with their edges staggered.

adhere to the lofted lines, work slowly and carefully. Make your bends close together and check the progressing shape after each effort. Try not to overbend for in attempting to rectify your error you could distort or kink the pipe, which will mean starting all over again with a new section. There is no need to use heat to aid your bending if you are using the right tools. Heating can cause a loss of strength and increase the danger of kinking.

Bending the pipe-frames can be carried out by one person, but it is always useful to have someone else around to give you a hand, particularly when it comes to checking the curves. To aid you in this respect a few short blocks of wood nailed to the loft floor on either side of the pipe-frame (see photograph) will hold it in place while you eye the accuracy of your bending. While the 'half frame' is in place mark on it the waterline and sheer line. Anything that won't rub off can be used; a drop of paint or even a strip of marking tape will do. As each pair of frames (port and starboard) is completed in this fashion label them clearly with their station number for ease of identification later. Store them out of the way until all the frames have been formed and you are ready to erect the framework.

Now to form the stem, keel and stern section. If the profile was not lofted at the same time as the transversals, this is the time to do it. Unless you can erase the existing lines on the lofting floor it might be an idea to give the surface a dab of paint so that you have something clear to work on. But leave the grid for you will obviously be needing that. Plotting the coordinates is carried out in the same way as for the station lines, only this time you will mark in the sheer, keel, sternpost, horn and rudder.

The pipe used for the stem will be shaped in the same way as the station frames. Some builders use a larger diameter pipe to provide a stronger 'backbone' for the structure. If possible use a single pipe length for the stem which hopefully will reach down as far as the forefoot. The construction from there aft to the stern can vary considerably. Often the stem pipe may be extended down the centre of the keel to link up with the base of the sternpost. This will give the hull extra rigidity. Alternatively the stem pipe will end at the forefoot, welded to the front end of the keel loop. Obviously whatever system you use is guided to some extent by the type of vessel you are building.

The way in which the keel is framed may also differ according to the hull design and the builder's preferences. In yachts which have a trough keel, a design which particularly lends itself to ferrocement construction, there are numerous ways in which the keel can be prepared. Naturally the lines for this section will have to be lofted. One method, often adopted, is to fabricate the keel loop using pipe (of the same diameter as the station frames) bent to the necessary shape and welded fore and aft, connecting with sternpost and stem. Welded across the 'loop' are trusses, at predetermined centres. Mild steel rod can be used for the trusses. High tensile rod can also be welded longitudinally along the keel trough, although some don't consider this necessary. As a variation, a keel backbone of pipe can be laid down the centre of the keel loop, with the trusses welded to the top of it and to the edges of the loop.

Some builders, clearly concerned with the possible damage that may be caused to a concrete keel by touching ground, which will happen at some time during the vessel's life, construct the keel using a steel base plate. The plate must be fairly thick; about ½ in. to ¾ in. would be adequate. This is trimmed to the shape required using a cutting torch. The plate is welded to the forefoot and the station frames along its length. In the actual construction the outer layers of mesh are laid round the outside of the keel covering the steel plate so that the implaced mortar held within the mesh matrix and covering it will give some protection from corrosion. These are but a few ideas used for the construction of the keel. No doubt there are many others!

18 Detail showing the trough keel of a 48 ft yacht. Note the large diameter pipe extending down the centre of the keel and providing a strong backbone to the structure. The pipe-frames have been welded firmly to the keel loop with solid rod braces holding them apart; rod stringers are not welded to the frames but held in place with wire ties. In this case the first layers of mesh have been applied to the outside of the reinforcement.

There are also a number of ways of constructing the sternpost. Perhaps the most frequently adopted technique is to use steam pipe. Working down from the horn piece, the pipe can be bent to the shape of the sternpost taking in the propeller aperture and carrying on down to the base of the keel. Provision has to be made for propeller shaft within the 'cut-out' and for the rudder stock at the top of the sternpost. If a single pipe section shapes the stern, this can be interrupted by a tube (somewhat larger in diameter than the propeller shaft or stern tube is likely to be) which will give access into the hull. This tube will form a joint to which the top and bottom pipes forming the sternpost can be welded. Similarly a tube can be welded to the top of the sternpost, taking the correct angle for the line of

19 Close-up of the top end of the stern post showing part of the cut-out for the propeller. The rod stringers are bent round the post, giving shape to the stern, and wired together. They should not be welded as this could cause distortion.

20 Plate steel is often used to form the stern post, with a pipe section welded in place to take the stern tube. The reinforcing rod stringers are then wrapped round the form to give body to the stern. Gudgeons from which to hang the rudder can be welded in place. It is a good idea to give them a protective coating of epoxy prior to plastering. Some builders have found it difficult to achieve proper penetration of mortar with so much steel round the stern and have used a simpler construction.

21 The stern post, a shaped section of $\frac{3}{4}$ in. plate steel, is set up in position and shored with timber. Note that the horn piece is already welded in place. Photos **15–21**: John Hodsdon.

the rudder shaft. Instead of a single pipe to shape the trailing edge of the keel two can be used, to fair the keel in gradually. Sometimes, if for example the builder is not too sure of the line of the propeller shaft from the proposed position of the engine to the propeller aperture, the tube will not be used. Instead a small box frame of steel flats is constructed to hold together the sternpost and give a square aperture. When the time eventually comes to fit the stern tube, presumably after the hull has been plastered, it is a relatively simple matter to align the stern tube and propeller shaft. Scrap pieces of mesh and rod can be stuffed round the stern tube and mortar injected into the cavity to bond in the tube and provide a sealant. This method has often been used here in Britain. The idea of imbedding the stern tube permanently in the concrete does highlight the possible problems later if for example the tube becomes damaged and has to be removed for repair or replacement. The adoption of a tube former to which the proper stern tube and shaft bearing can be fastened does offer a safeguard in this respect.

Another method of constructing the sternpost makes use of plate steel, about $\frac{3}{4}$ in. thick. The shape of the stern is lofted in the usual way and the pattern taken off on tracing paper. From this a template is cut from a sheet of plywood. The steel plate can then be patterned and shaped. Allowance must again be made for the propeller shaft and a tube can be welded to the plate (see photographs). The pipe horn piece can be welded to the top of the sternpost. Short lengths of pipe can be welded round the trailing edge if need be to aid the fairing of the stringers round the stern. Rudder fastenings, as in the photograph, can also be attached. But with so much steel in the area this method gives rise to some misgivings that it will be difficult to ensure an adequate penetration of the mortar round the reinforcement. There is a distinct danger of voids being created.

Prefabricated rod frame

The lofting and shaping of rods for fabricated frames is carried out in much the same way as for pipes. One set of rods will reflect the curvature of the inside surface of the hull shell. The other set will follow a parallel curve some three inches further inboard toward the centreline. Richard Hartley, the New Zealand builder who first developed this technique, suggests $\frac{5}{16}$ in. mild steel rod for the frames as suitable for vessels up to at least 70 ft.

Begin by bending the outer rods to the required station shape, marking in the waterline and check line. Similar bending equipment to that used for pipe can be used, but is slightly modified of course. Do not use heat for it will make the rods brittle and increase the danger of kinking. Now bend the inside rods to the pattern. Throughout the operation work slowly and carefully making frequent checks to ensure that you are following the loft line. It is usual practice to incorporate the deck beams in the structure, so you can also prepare the rods for them, shaping them on the prepared beam loft. Where there is only a narrow side deck (along the coach roof) and the beams do not extend right across the boat, the rods forming the station frames can be sized and shaped to include them.

To fabricate the frames set the outer rods down on the full sized pattern, and tack weld the ends. This applies to those frames which join the stern pipe or stem. Where the stations come down to the keel allowance must be made for the width, and a short section of mild steel rod of the right length can join the starboard and port outer frame. The inner rods can be placed in position and similarly welded together. Remember that during the welding considerable heat will be generated which could cause some distortion of the rods. Ensure that during this process the rods are not bent out of alignment. Now for the trusses. These are welded between the parallel rods. A continuous length of $\frac{5}{16}$ in. rod is looped on a jig and shaped to the required form. Do not be disheartened if your first two or three attempts at fabricating these frames end in dismal failure. A certain amount of expertise is needed and this will only be obtained with practice. Naturally there are variations in the way these frames may be constructed. A certain versatility and inventiveness would not be remiss in their fabrication.

As each frame is completed label its position clearly so that there is no confusion later. This method of prefabrication using rods is not adopted for the formation of the stem, horn piece and such places like the stern frame, where a web will not be needed. Pipe is used instead, in the same way as for the pipe-frame technique.

The prefabricated frames, besides virtually eliminating the danger of deflection, provide flanges or webs convenient for the attachment of bulkheads and other interior fittings. However, it is unlikely that you are going to be able to make use of the webs of each station frame; they will merely be useless appendages. Why not combine the prefab frame and the pipe-frame techniques, placing pipe frames where no webs are required and using prefabricated frames where they are? The decision should be taken at this stage, when the frames are being lofted and shaped.

Wooden frame

As an alternative to pipe or solid steel rod, wooden frames can be constructed to produce the hull form. They will, of course, be removed once the hull has been plastered and the mortar is no longer in a green state. It is not really a type of framing that I would recommend in place of either pipe-frames or the prefabricated frame. No doubt they are easy to construct and will provide a rigid shape, but they do present certain disadvantages. In the first place they are temporary structures, only there to provide the initial shape of the hull. The two other types of frame will remain as part of the hull structure and, despite some thoughts to the contrary, will undoubtedly add to the overall strength and rigidity of the hull. Wooden frames do not provide a means, like a flange or web, for the attachment of bulkheads, etc. These have to be fabricated in addition to the frames. However, despite these weaknesses wood framing is a suitable alternative to the other two if those materials are not readily available, or you are concerned about your ability to manipulate pipe and rod and to weld it.

The construction of wooden frames is relatively simple and proceeds along the lines for the conventional framing in a wooden hull. Spruce planks or some other soft wood can be used for the frames. They should be at least an inch thick so as to provide a reasonable surface for nailing or stapling on the rod stringers.

Do not forget when lofting your lines that the measurements, unless otherwise specified, indicate the outer surface of the hull. The thickness of the shell should be subtracted to give you the curve for the station frames. With the coordinates plotted and the lines drawn in, trace out the pattern for the half station. A good stiff tracing paper will serve, strips being taped or glued together to cover the whole half station. Give yourself a bit of space to play with, so don't use a strip of paper which is too narrow. With a pair of scissors, cut along the station line to produce a pattern. In drawing out the pattern mark in the waterline and sheer line.

Now you can either use this flimsy pattern to make your frames or a further template constructed out of a thin plywood. Quarter-inch ply would be more than suitable. A bandsaw can be used to cut out the shape of the half station. With the spruce planks the half station can be shaped. Where two or more pieces are needed to complete the curve, these can be fastened together with butt blocks, glued and screwed. Keep all these blocks on one side, and out of the way of any likely bulkhead locations. The webs for the bulkheads will be fabricated while the wooden frames are still in the hull. With a half station completed it can be used as a template for the other half.

The two half stations are now laid on the lofting floor so that their edges just touch along the centreline. Square up each station to the waterline, and fasten them together with a butt block, cleat or whatever you may wish, and label the joining strip. Now a frame spreader will be needed to hold the frames firm. A 1 in. timber beam will suffice. This can possibly be set along the waterline, and further timber braces taken from it to strengthen the structure. The frame spreader, if it is along a known height line, can be used as an indicator for levelling the frame.

Timber forms can be used to provide the shape for the stem and the stern, but are not recommended. Use instead the conventional water pipe and continue the pipe along the centreline of the keel. This backbone connecting stem to stern will add considerably to the rigidity of the structure and will enable the hull to withstand far better the compressive, tensile and impact stresses exerted on it. The pipe will also provide a convenient and firm anchor on which to attach chainplates for the forestay, a chain roller or fairlead. Extended above deck level it can act as a brace to a bulwark if one is constructed.

Setting up

With the frames bent to shape you can now set about fabricating and erecting them on site to provide the mould for the rod and mesh armature. The way in which this part of the construction is carried out is to some extent governed by the facilities on site. A support structure will be needed from which to 'hang' the framework. If the construction is to take place in a shed the rafters, if exposed, can be put to good purpose. If this facility is not available some sort of structure will have to be built. The best way to build the hull is upright. It will enable you to get at the inside of the hull more easily, and you won't have the problem of turning the hull right side up.

One of the simplest and most efficient ways of building a support structure is to make use of scaffold piping, erected to form a skeleton framework of uprights and rafters like a trellis. If properly constructed with sufficient braces to hold it firm, the structure should be more than strong enough to hold steady the hull. The cross members should be so placed as to have one above each station frame. The frames will be supported by them until the hull has been plastered and cured.

The trellis will need to be assisted in the support of the hull, for it won't be able to do the job on its own. Props will be required as well, strategically placed at the turn of the bilge and at any point where there is a danger of the framework sagging under the weight of mortar, or deflections being caused during the placement of the mortar. This is particularly relevant if the pipe-frame technique is adopted. Particular attention must be paid to shoring the hull.

Timber uprights and beams constructed to form a trellis will serve as well as scaffold piping. But use timber which will provide a rigid structure. Bear in mind that you will require a raised platform or scaffold round the hull from which to work. So make allowance between the uprights of your support structure and the outside of the hull. The framework can also be used to good purpose, particularly if it is out in the open, to support a roof of sorts.

During the lofting of the station lines you will remember that for the formation of the grid three essential lines were marked out. These were the centreline, the waterline (l.w.l.) and the baseline or datum. The co-ordinates were plotted in relation to the baseline and centreline. These lines are once again going to be employed during the erection and fairing of the hull framework. A position line down the middle of the building bay will constitute the centreline over which you will build the hull. The baseline will lie directly above the centre on a horizontal plane. All measurements can be taken from this line when setting up the station frames. The baseline can be formed out of a straight section of water pipe (or sections welded together to make up the necessary length), a length of angle iron, or even a timber beam. Alternatively stiff wire (galvanised fencing wire is suitable) can be stretched tightly down the middle of the building bay at the desired height. The cross beams or rafters of your support framework could also be placed on the same plane as the baseline, with the top edges level with the baseline.

A solid foundation will be needed on which to rest the keel. Any settlement or movement of the hull, which could be caused by the collapse or subsidence of the base supporting the hull, especially while the mortar is still in a green state, could distort the fairness and create weaknesses within the structure. If you are building in a shed or warehouse which has a wooden or concrete floor, this problem will not trouble you. But for those who have, perhaps, to make do with their back yards or other sites in the open, a solid base will have to be provided. Railway sleepers (about three or four would do) laid crosswise in the area of the keel are often used. Even if you are working on a concrete floor some sort of shoring for the keel will be needed. A 2–3 in. thick plank, the width of the keel, and blocked up or wedged to the angle of the profile, should be strong enough for hulls up to 70 ft.

There is no set order in which the framework of the hull should be assembled. It is purely a matter of personal preference. Some builders prefer to start with the profile (stem and keel first followed by the sternpost, horn and transom). If the keel consists of a single pipe, then it and the stem can be welded together on the loft floor so that a check can be made to see that the heat does not cause any distortion. In this position mark the positions of station frames plus the height of the stemhead and deck.

This part-fabrication can now be placed in position in the building bay, held in place by props. A soft wire tie could hold the stem to the rigid baseline, the keel shored up temporarily. The sternpost, with horn piece already welded in place (see photograph) can now be tacked to the keel and likewise held in place with props. A plumb bob or measuring batten and spirit level can be used to check the height and line of the stern, taking measurements off the baseline.

As a slight variation, where the keel has a wide cross section and has been constructed in the manner discussed earlier in this chapter, the keel frame can be laid along the centreline, and then the stem and sternpost temporarily tacked in place. These tack welds can later be broken if necessary to allow for further adjustment of the frames. It is now the turn of the station frames to be erected in position, but first each pair of frames should be placed back on the loft floor and their shapes carefully checked to see that they are still true. The half-breadths are then placed opposite each other on the floor, as they would appear in their rightful positions, as part of the hull framework. Their bases can then be tack-welded together, to form a rough triangle if the frames connect with the stem or horn piece; or short braces or spacers can be used to hold them at the base if the stations connect with the keel section. If pipe-frames are to be used for the deck beams, then these shaped beams can similarly be tacked to the frames, on their marks. This job can also be left till later when the frames have been positioned. Remembering the inherent weakness of the pipe-frames, further braces or frame spreaders can be welded between the frames, perhaps at the waterline or the turn of the bilge. At this stage these must only be tacked, in case adjustments have to be made later during the fairing, which will surely be the case. To aid the hanging of each frame and the fairing, a further pipe can be employed in the assembly. This is tacked vertically between the frames forming the centreline, from the keel (welded to the V-joint or bottom spacer, as the case demands) up to the centre of the deck beam and beyond the overhead baseline. When each frame is erected in place this vertical can be tied to the baseline (if it is rigid) with soft wire ties, or welded in place once the frame has been adjusted to its correct height.

Each frame assembly can now be erected on the site. Work systematically from one end to the other. A bit of ingenuity would come in handy to hang the frames. One idea is to adjust the height of the frame with the aid of the vertical upright, wiring it in position. The cross bearers or rafters of the support scaffold can now be brought in to hold each frame as it is hung. A possible further addition, which has been frequently adopted, is to link the upper ends of the frames with wire hangings taken over a central ridge pole. This pole could be the baseline, if it is high and rigid enough, or a rafter directly above it. The height of the frames to the baseline or datum, from which all measurements are taken, can be adjusted by lengthening or shortening the wire as needed. Steps must now be taken to prevent undue movement of the frames. If they extend beyond the cross bearers of the scaffold they can be attached to them. Props must also be used, particularly when fairing has taken place.

Long 1×2 in. timber sections can be used for these props. An effective method of holding them to the frames is to screw or bolt a short strip of metal or angle iron to the prop, using it as a clamp to grip the pipe-frame. The clamp can be loosened to allow the frame to be adjusted. The prop can be long enough to reach up as far as the deck beams where further clamps can be fixed (see photograph).

With all the frames now in place the stem, keel and stern section can be welded on. Tack rather than put in a strong weld for the framework still has to be faired up. You can start by ensuring that each pair of frames is still set at the right height to the baseline and is standing vertical. A spirit level and measuring batten will be of assistance. But before getting your teeth into the job of fairing, the layers of mesh which are going to cover the underside of the keel should be spread. Three or four layers of mesh, depending on the type of fabric and scantlings, is the usual quantity. Often builders may increase the percentage of steel at the keel, bearing in mind that this area of the hull will have to bear the brunt of any damage which may occur should the vessel take the ground. Note, however, that increasing the concentration of mesh in this area may compound the problem of getting good mortar penetration. I cannot stress too strongly that this must be achieved.

Besides the mesh cover builders sometimes construct a simple rod keel framework consisting of longitudinals and a series of braces welded up the length of the keel. This is to further strengthen the keel structure and to continue the even distribution of the high tensile rod stringers round the hull. The braces can continue some distance up the side of the hull, perhaps for a foot or so, to link up with the transversals. Where transversals are not used, or only to a limited degree, the braces should extend only to the extremities of the keel.

The mesh layers used for the keel should be wide enough to lap the sides of the hull. Stagger the edges so that joints do not occur along the same line. Besides creating a potentially weak spot, the joints could cause a buildup of mesh resulting in a ridge. The mesh layers should also be at least as long as the keel, but preferably longer to provide an overlap. The way in which the keel fabric is laid out is illustrated in the accompanying photographs.

22 The framework of a ferrocement hull being completed in Mill Valley, California. In this construction $\frac{1}{2}$ in. internal diameter pipes were used to make up the frames. The reinforcing rod stringers, $\frac{3}{8}$ in. diameter, can be seen being wired in place. This particular builder was fortunate in having a shed in which to build. Vertical supports have been dropped from the rafters, attached to the pipe beams of the hull to support the structure.

23 An early stage in the construction of a 63 ft ferrocement yacht built in Brentford, West London by Chris Cole. The pipe framework has been set up and is almost ready to take the high tensile rod stringers. Note the technique used for propping the frames. Photo: Chris Cole.

24 Frames, stem and keel section are all in position and tack welded temporarily, and held firmly from above by the cross bearers of the support structure. Fairing can now begin. Photo: J. Hodsdon.

Now to the fairing of the framework. Shore up the keel with timbers suitably wedged. Remember that most of the weight of the plastered hull will have to be supported by this shoring, so make it strong and immovable. Once again check that all the frames are level and at the right height in relation to the baseline. It is quite likely that while fiddling about with the mesh and the shoring, the structure may have sagged, the frames moving out of true. With a spirit level check that the marks on each frame indicating the waterline are still on the same plane.

Thin wooden battens will now be needed to fair in the frames to ensure a smooth and continuous curve from bow to stern. At least three battens are required on each side, stretched the length of the hull. One can be placed at deck level to indicate the sheer. If you had the forethought to indicate the sheer line on the frames when you fabricated them, this job should be relatively easy. The second batten can be laid along the waterline, and the third beneath the turn of the bilge. The battens will take up the natural curve of the hull. Frames which are out of true or shaped incorrectly can be picked up. Where battens don't touch a frame, do not attempt to work the batten in to meet it. The frame must conform to the batten not the batten to it. Where a frame is out of true it can be adjusted. This might necessitate breaking the weld and re-positioning the frame. In extreme cases you could be forced to remove the complete station and reshape it. But be careful that you don't distort or kink the pipe in the process. It might be more expedient to fabricate a completely new frame. This would probably be the case where prefabricated rod station frames are being used. But it is to be hoped that the original shaping of the frame was not inaccurate enough to warrant constructing a replacement. Where wooden frames may be slightly out of alignment, the support structure may need some adjustment. These frames will be wired to the pipe backbone and wooden verticals can be used to suspend the frame from the cross bearers. Fore and aft braces will need to be positioned between the frames to give the structure further rigidity. All of these appendages may need altering to get the hull mould properly faired.

Do not rush this stage of the construction. You will need to be patient to get good results. This is only the beginning of the fairing. After the application of the reinforcing rod and mesh you will need to spend a couple of days ensuring that the hull is completely fair. During both stages, particularly where problems might crop up, get a professional builder or architect along to advise you. His experienced eye is more likely to spot mistakes which, in your inexperience, you may have missed.

The preliminary fairing of the framework is now virtually complete. But before the reinforcement (rod and mesh) can be applied to the mould, the deck line must be fixed. Steam pipe can be bent to the sheer from stem to stern and welded to the inside of the frames, on the marks fixed while the frames were still on the loft floor. This pipe can be positioned just below the deck beams, if they have been erected at this time. Instead of pipe, a rod stringer can be used, wired onto the outside of the frames. This will reduce the possibility of the structure moving or distorting while the rod and mesh are bent on.

ferrocement 8

REINFORCEMENT

Once the framework has been properly erected and faired up the reinforcement can be attached to the ends and frames. We with the stringers or horizontals.

Rod and mesh placement

Rod placement

High tensile rod is recommended, but mild steel rod can be used. However, understand that by using the latter considerable reserve strength is lost. In present-day practice $\frac{1}{4}$ in. diameter rod is the most common dimension for the longitudinals of hulls between 30 ft and 70 ft l.o.a. In selecting this rod ensure that it is straight and without kinks. Quantity and placement of the rod is obviously governed by the structural design, but in general the rods are spaced at about 2 in. to 4 in. centres round the hull.

In bending on the rods begin at the stem and work round to the stern. The rods can either be hooked round the stem or welded to it. In bending the rod do not use heat, as this could distort it and will almost certainly weaken it. Do not overbend the material for this might later split the concrete or fracture the rod. Following BS 4466 the radius of the bend, for high tensile steel, should not be less than six bar diameters. With mild steel the radius should not be less than four diameters. If the rods are to be bent round the stem they must be wired on. Soft baling wire can be used. While it is not recommended practice to weld high tensile steel, this can be ignored when attaching the horizontals to the stem. Welding to the stem is much easier than wiring. But do not weld the horizontals to the frames: this will cause distortion and loss of strength. Stick to wiring them in place. It is a good idea when laying up the rods to attach them in pairs, one port and one starboard, particularly at the beginning. It will enable you to keep the whole structure in balance and decrease the chances of any of the frames being forced out of true. Bring the rods back as tightly as possible along a horizontal line and lap them round the stern or hook them to the transom frame. If possible use full length high tensile rods for the stringers. Not only will you get the full strength of the rod, but it will also enable you to work the rods into smooth curves round the framework. Unfortunately the rebar can normally only be obtained in up to 40 ft lengths. For larger vessels you will obviously need two lengths of rod to make up a single stringer. Overlap the rods by at least a foot. Do not weld the ends together but tie them. Welding will distort their shape, create kinks, and weaken them. Baling wire wrapped firmly round them will suffice. Try to stagger the joins so that you don't get a build-up of steel in only one area. Cutting or cropping the rebar to size can be accomplished with a powerful long-armed bolt

25 High tensile rod stringers wrapped around the pipe stem and tied in place. Note the overlap, which should be 9–12 in.

cutter with hardened jaws. These are usually available from most hardware stores. Another method is to use a gas cutting torch; or heating the bar to weaken it and then using a hacksaw. But be careful how you use the heat, to avoid weakening the metal which is left, or other parts of the structure. A minimum of heat should be used, and only on the metal to be cut.

In bending on the stringers work from the keel up to the deck line. With every stringer check that you have a smooth curve taking the shape of the hull. Keep them as tight as possible. If you have resorted to using mild steel rod, which I repeat is not recommended, you may find that between each frame you will get the stringers drooping in scallops. Drawing the rods round the frames could cause flats, as they might kink to take up a broken curve. One possible way of reducing this problem is to use more frames than usual, but it would not totally eliminate the flats.

The way in which you place the stringers will to some extent determine the eventual finish of the hull. Make sure that the stringers take gentle curves. Where kinks occur, you will find that the hull will develop a hollow or a bulge. This will have to be built up later with mortar, an unsatisfactory solution. While initially the appearance may be acceptable, the first time you severely bump the hull, coming alongside somewhere, there is a good chance the remedial plaster 'build-up' will flake off or crack.

If you are going to use transverse rods in your structure now is the time to apply them. Many ferrocement boats have been built without them, containing only the fore and aft members. It is a matter governed by the construction design and the strength figures of representative test panels. Some architects don't bother to work out their own strength calculations and instead base their construction design on previous ferrocement vessels which have proved sound in structure. The strength requirements of a hull are largely guesswork, for no precise values have yet been established for calculating the magnitude of the stresses imposed on a hull which is 'operational'.

So, builders tend to hold conflicting views about the importance of incorporating athwartship rods in a ferrocement hull. New Zealand builder Richard Hartley, for example, considers that transverse rods are really superfluous in boats under 40 ft l.o.a. He feels that the actual strength gain is not warranted by the subsequent increase in weight. For yachts above 40 ft he does, however, advocate transversals, or rather diagonal rods, set at an angle of 45 to 60 degrees to the vertical. Tests he carried out on representative ferrocement panels indicated that the inclusion of transverse rods placed uniformly throughout the structure increased the weight by 40 per cent. The strength gain was only $7\frac{1}{2}$ per cent.

26 View of stem pipe with stringers wrapped round. Here it is usually good practice to weld the horizontals to the stem. Photo: J. Hodsdon.

27 Stringer being wired in position round the pipe frames of a deep keel yacht. Note the method used here to support the framework. Wooden cross bearers are set up at a convenient predetermined level and each pair of pipe frames wired firmly to them.

Most other builders, particularly in this country, feel that some form of transverse reinforcement inside the mesh layers is advisable. While it may not be as necessary in a shallow draught hull, such as a V-bottomed motor cruiser, it can be important in a yacht with a heavy ballast keel. Such keels will need something more than longitudinals, mesh and mortar to hang on to. Transversals or diagonals would provide this anchor. Their extra strength is a good safeguard against serious damage.

To illustrate the type of damage which may occur through the exclusion of transversals I quote two examples. In the first, the shores holding up a deep-keeled ferrocement yacht apparently collapsed causing the vessel to topple heavily onto its side. It seems that there was no actual distortion of the hull as a result of this accident, but long parallel cracks were discovered at the top of the deadrise. There were no vertical cracks at all. In the second incident a similar yacht had gone hard aground and was left dry when the tide turned. Stresses on the keel similarly caused horizontal cracks in the keel. Both vessels had been constructed without the inclusion of athwartships rods.

In common with present practice the transversals are spaced further apart than the stringers, often at about 5 in. or 6 in. centres. Other builders reduce the concentration of transversals further, by limiting them to specific areas. One British builder of a 42-footer incorporated three transversals between each frame at about 3 in. centres. Understand also that where prefabricated frames and pipe frames are used, they will also be acting as transversals to connect the keel to the main structure of the hull. It is also recommended that transversals be introduced to the area surrounding the proposed position of the chainplates.

These athwartships rods are often of a much lighter gauge than the stringers. Diameters range from $\frac{1}{8}$ in. to $\frac{3}{8}$ in. In applying the verticals or the diagonals work from the keel, either welding the rods to the keel frame, or tying them to the rods which were introduced under the keel. Overlap by at least a foot. Lead the rods up and around the turn of the bilge to the deck, tying them to each horizontal. Ordinary soft baling wire can be used for this. Another idea is to use what is called a Stabil binder. This is a shaped binder produced specially for the purpose of tying reinforcement together, and is manufactured in Britain by Huntley & Sparks Ltd, of Mitcham, Surrey. One builder who used these binders reports that they considerably reduced the tying operation. They also appear to have improved the 'resilience of the hull basket'. No doubt similar ties are obtainable in other countries under different trade names for the construction industry.

If a ferrocement deck is to be built the transversals can be extended to include the width of the deck. Alternatively, if a web is to be prepared on which to fasten a wooden deck (discussed later in the text) the rods can be bent to provide starters. Further longitudinals can be used for the deck, fastened on top of the transversals from stem to stern. A scaffold or working platform of sorts, placed inside the hull, will be of use when applying the deck reinforcement. Many builders find that it is more convenient to fabricate the deck after the hull itself has been plastered. The deck reinforcement could get in the way when pouring the mortar.

As you might well realise, the inclusion of transverse rods will increase the cross section of the ferrocement shell, thereby increasing the weight. Weight can be a critical factor in a yacht and too much above the load

28 If a ferrocement coach roof is to be incorporated it can be framed and reinforced with rod in the same manner as the hull. Here the after end of a coach house is illustrated. Rods are drawn back to the pipe frame and either bent round it or bent and welded in place as in the photograph.

29 The reinforcement for deck and bulwarks in place. Notice that the transverse rods of the hull are extended up to brace the bulwarks. A large diameter rod has been bent round to provide the deck sheer. Pipe is more frequently used for this job.

waterline could affect the boat's stability and sail-carrying capabilities. In an effort to reduce this weight, one builder suggests halting the transversals at the waterline, or just below it, providing a compromise. The thickness and hence hull weight below the waterline is not too critical a factor and can be linked with the static ballast. The shell above the water line would therefore be considerably thinner, and much lighter.

After all the rods, both horizontals and verticals, have been laid on, it is once again necessary to spend some time checking the trueness of the hull form. A 6 ft wooden batten would come in useful here. Note that already the rod reinforcement will be exerting considerable loads on the frames. Where the rods may have been bent out of alignment, either inwards or outwards, try to correct them now. Attempting to adjust them once the mesh has been laid will be most difficult. A rubber mallet or hammer can be used to bang them back into place. If any of the rods have sagged in any way it might be necessary to rewire them. Pay careful attention to the fairing round the deadrise. The transversals tied to the horizontals will be exerting a considerable outboard pull which could cause a flattening of the curve or even a bulge. This can be rectified by tightening the horizontals. At this stage it might also be prudent to go over your cross braces within the hull. Make sure you have enough to keep the framework rigid. At this time you can also set about placing a proper scaffold within the hull, if you haven't done it already. You will need a working platform inside for moulding on the mesh. You could now also think about any rigging fittings which you may wish to weld to the main frame structure of the hull. This could include chainplates and perhaps the chain roller or fairlead for the bow. You could also consider the fabrication of the toerail or bulwarks.

Mesh placement

The steel content in a ferrocement structure must be as high as practicable and spread uniformly throughout the form. In specific terms this means that the steel content must not be less than 24 to 36 lbs per cu. ft. This will allow for the provision of a steel mesh content of 2 to 3 lbs per sq. ft for a 1 in. cross section of hull. Higher steel concentrations have been used for a ferrocement hull, but the amateur builder can barely reach the lower recommended ratios of steel to concrete, even with careful tying and wiring of the mesh.

There are a variety of mesh fabrications available, as discussed in the chapter on construction. The two most common types are the diamond shaped chicken wire or bird netting, and welded wire fabric. Galvanising is

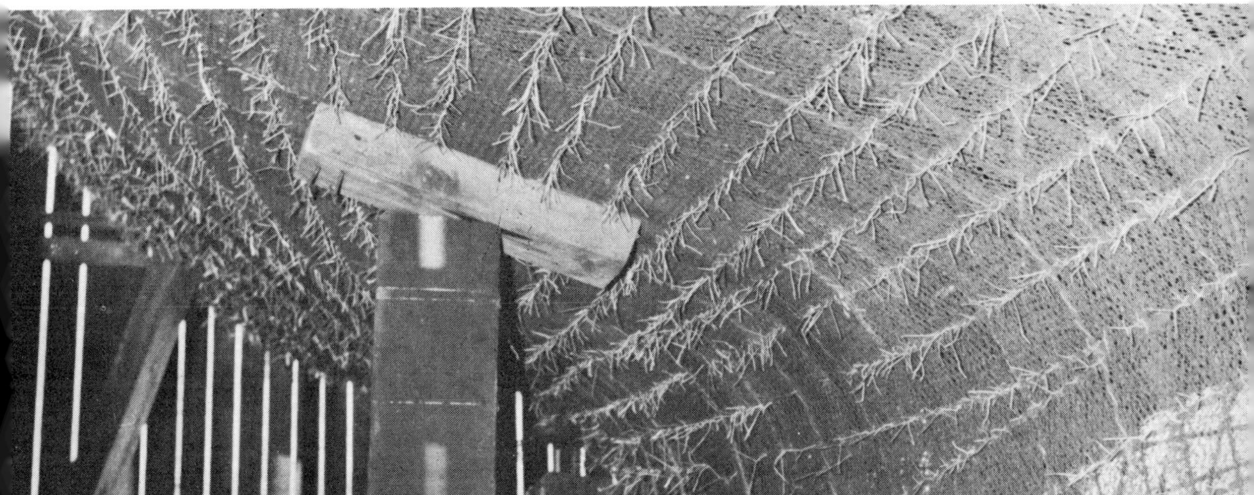

30 This photograph illustrates in no uncertain terms the number of wire ties that are likely to be necessary to draw the mesh in tightly against the rod reinforcement. The picture is of a 40 ft hull, redesigned from a timber-built Perryman, constructed in Kilmarnock, Scotland. It would have been far better for the builder in this instance to have tied the wires on the inside of the hull rather than the outside. All the protruding ends will have to be pushed back into the reinforcement before plastering can take place, and certain slight imperfections will result which will surely show up in the final appearance. Photo: A. L. Edge.

recommended, particularly after fabrication. Remember that only a thin layer of dense mortar will cover the external surface of the mesh; the galvanising will provide added protection to the steel. Of the two, welded mesh is regarded as far superior. This is not to say that chicken wire has not been successfully used in ferrocement construction. It has been, and many professional builders still do use it in preference to the stronger fabrication. This more because of cost and availability than any other consideration.

The steel in ferrocement provides the tensile strength, while mesh is there to spread the reinforcement uniformly and to take the stresses induced during impact or twisting. Correctly, the structure must be flexible and only yield under unusual loading. An American naval architect, Jay R. Benford, NA, of Friday Harbour, Wash. discusses the comparative strengths of bird netting and weldmesh in an article in *National Fisherman* (April 1971). He refers to the results of bending tests using sample panels of weldmesh and bird mesh. The sample with layers of weldmesh broke at twice the loading of that using chicken mesh. The conclusion to be drawn from this is obvious.

The type of bird netting commonly recommended by boatbuilders is 18 to 22 gauge with openings of $\frac{1}{2}$ in. to 1 in. The $\frac{1}{2}$ in. 22 gauge is the most frequently used. To make up the necessary weight of mesh at least eight layers are wrapped on. The weldmesh is a much stiffer and heavier material, and only six layers are needed to obtain the necessary 2 to 3 lbs per sq. ft for a 1 in. thick hull, required for boats up to at least 70 ft l.o.a. The weldmesh is 18 to 19 gauge with square openings of $\frac{1}{2}$ in. to $\frac{3}{4}$ in. Often combinations of the two types of mesh are used, with the weldmesh usually on the outer surfaces of the hull membrane because its greater rigidity helps it to hold its shape when plastering and fair up irregularities in the inner layers of mesh. However, always keep in mind the importance of maintaining a high concentration of steel throughout the shell.

Applying the mesh and tying it into a tight matrix is the most tedious and time consuming stage in the whole construction programme. It is a painstaking job which requires a great deal of patience. It is also a very important phase. The mesh must be laid on smoothly and evenly; stretched tightly over the framework without bulges anywhere on the hull, which would create unfair hull lines. Looseness of the mesh could also aggravate the problem of mortar penetration, and could also affect the finished weight of the hull.

The mesh is laid on sandwiching the rod stringers and transversals. In the case of the eight layers of bird netting, four layers are placed on the inside and four outside. Similarly, the six layers of weldmesh are laid on equally with three on either side of the rod reinforcing. It does not matter from which side you start laying up, but putting on the inside layers first seems to be more popular.

Where wooden frames are used to provide the mould, the mesh layers in the area round the frame, on the inside, are applied before the stringers and transversals are placed, for the frames are only temporary constructions and will be removed after the hull has cured. Before putting up the mesh, treat the edge of the frames with some sealant. A thin polythene or plastic sheet stapled to the wood is often used. It acts as a vapour barrier and will prevent the wood absorbing moisture from the wet mortar.

The best method of applying the mesh is to take a rough measurement from the deck line down to the keel and cut off the desired length. Ordinary wire cutters will do the job admirably. If you have a ferrocement deck planned and set up, extend the mesh layer to cover the width of the deck. And anyway level some overlap for a web flange on which the deck may later be fastened. Allow at least 9 in. to 1 ft overlap at the bottom where the already fixed keel mesh comes. Begin by hanging the mesh from the sheer, then tuck it in at the keel and work up, flattening it against the frame. This will prevent the formation of 'bubbles'. Staple each layer loosely to the frame. Once the longitudinals are properly fastened in place the mesh can be stretched and smoothed in properly. Sometimes it may be easier to fold the mesh giving it a 1 ft 6 in. width which is more manageable. Two doubled layers will give the necessary thickness on one side of the rod reinforcement. Where pipe framework is being employed with the idea of later removal the same system will have to be adopted as applies to the wooden frames. The mesh must first be fastened up in the region of the frames and then the rods fixed in place. Soft wire ties can be used to bind the mesh and rods to the frame.

In the case of fabricated rod frames, both sides of the projection must first be covered with mesh. Fewer layers may be used for them as they will not be subjected to stresses of the same magnitude as the hull itself. If bird netting is to be used three layers on each side will provide sufficient concentration of steel. Place each layer of mesh firmly over the protruding frame flat against it and allow at least a foot overlap onto the hull. The normal width of the mesh will cover the area. Make certain at the angle alongside the stringers that the mesh is packed in tightly so that no mesh-free gaps are formed.

With the frames and the areas directly surrounding them meshed in on the inside you can continue covering the rest of the hull. Measure off lengths of mesh from sheerline to keel. The mesh is applied between the frames and parallel to them. Where successive layers are applied overlap the joints of the previous layer. Tie the mesh to each stringer as you mould it to the curve of the hull. The pressure of your hand should be sufficient at this stage to force the fabric into place. Work from the keel up. It is not necessary to concentrate the tying of the mesh as each layer is put down. Just enough ties, say at 6 in. intervals along each stringer, will suffice. Only when the reinforcing rod has been adequately clothed on both sides can the 'knitting' of the steel membrane be carried out in earnest.

It is not necessary to put down every layer of mesh parallel to the frames. It is just easier this way, particularly inside the hull. In practice it is a good idea to place the mesh longitudinally as well, to better absorb stresses imposed from all directions. The mesh can even be laid-up diagonally. This method is better suited to the outside of the hull. You will find that weldmesh, being a fairly stiff material, will take the shape of the hull better if wrapped on diagonally.

In applying the mesh fasten it securely at one end and then stretch it tautly over the hull. When applying the outside mesh ensure that it is long enough to properly lap the mesh at the keel. At the stern, transom and forefoot wrap the mesh firmly round edges, flattening all corners as much as possible. Unfortunately there does not appear to be any easy answer, or quick one, to the problem of tying on the mesh. One idea is to use long pieces of soft wire, sewing the mesh to the rod stringers. Although a bit quicker than any other method it is not recommended for you cannot pull the mesh up really tight.

The most common method is to use short wire ties bent in a U-shape through the mesh and twisted tight, pulling the layers in towards the centre. A similar idea used in Britain is industrial staples. These are galvanised and $1\frac{1}{2}$–2 in. in length, a convenient size for the desired hull thickness. Various instruments are adopted for the twisting and tying operation. Pliers will do the trick. Another is to use a pair of wire cutters: grip the tie between the jaws, twist up tightly and then clamp shut. This will crop off the excess length of tie. A further system, suggested by Hartleys and Ferro-cement Ltd, the New Zealand concerns, makes use of a pair of Vice Grips to which a bag tie is welded in a direct line with the jaws. The Vice Grips lock on the wire tie and then spin round to twist the wire.

Do not attempt to pull the opposite mesh layers too close between the rod reinforcement as this is likely to cause hollows and corrugations in the hull. It is sometimes a good idea to have two men working, one on the inside and one outside. Apply the tie from the outside pushing it through to be twisted from inside. Bend the ends flush with the surface of the membrane. If possible don't bind the mesh the other way around with the ties facing outboard, as only a very thin layer of mortar will be covering the wire armature and any small protruberances, like wire ties which haven't been knocked flush, will cause irregularities in the finish of the hull. They could be unsightly too, causing rust spots. The finish on the interior surface is not as important.

Initially the wire ties can be fixed at about 6 in. centres all along the stringers. Then as you start fairing the hull, once more, you can increase the concentration. The more there are the better, for it is important that the membrane is as taut and close-knit as possible. Always work from the keel upwards. This will allow any so-called 'bubbles' to be forced out at the top. Where you get a ridge forming, it might be necessary to cut into the mesh to smooth out the fold. But if you have been careful and worked slowly from one side to another this problem should not arise. For a vessel between 45 ft and 70 ft in length, the steel membrane should be about $\frac{5}{8}$ in. thick when you have finished compacting it and putting in wire ties.

Spend a few days after the mesh has been laid ensuring that the structure is as fair and as stiff as possible. During the placement of the mortar some force will be applied directly to the membrane. If it is at all baggy or loose deflections will occur which are not going to do much for the finished product. Where you do find slight bumps or hollows in the mesh these can be banged out with a rubber mallet or hammer. Take a wooden batten round the hull to help you check the fairness.

There is another variation to the fabricating of the hull armature which involves alternating the rod and mesh layup. Samson Marine Design Enterprises are experimenting with it, and tentatively labelled their technique the 'Sam-san' method. Their method incorporates 12 layers of 21 gauge chicken wire with 1 in. openings, together with the conventional number of $\frac{1}{4}$ in. high tensile rods placed horizontally and vertically. Four layers of mesh are first applied to the wood mould, followed by either the horizontal or transverse rods. Then four more layers of mesh are wired on; the second layer of rods; and finally the last four layers of mesh.

According to a report from Samson Marine, comparative strength tests carried out with sample panels formed in this way and in the conventional manner indicated that the 'Sam-san' technique exhibited a greater end strength. It was also easier to establish a good mortar penetration, while material costs were much less. At the time of writing the method has not been completely proven, but there does not appear to be any obvious reason for turning it down.

A direct argument for the technique would be that there is a greater and more uniform distribution of the steel reinforcement which, in theory, should improve the load bearing capabilities of the structure. With the conventional method the core of a ferrocement shell consists only of the rod reinforcement. By penetrating this area with mesh, which the Sam-san technique does, the core should be further strengthened. Samson Marine constructed a 37 ft workboat using this method and observed that they have not detected any more micro-cracking in the surface than one would normally expect if a finer mesh had been adopted. They reason that the presence of the middle layers could be a means of preventing, or at least reducing the possible cracking of the surface mortar during curing. I might add that the micro-fractures occur in many amateur-built ferrocement vessels. These cracks do not reduce the strength of the material to any extent, and they do tend to heal up, through a process described as autogenous healing.

Before any sort of plastering can take place provision must be made for the fixing of internal fittings like bulkheads, fuel and water tanks, engine beds and box, floors—in fact anything which may form a permanent attachment to the hull. Most of these attachments will require starter rods (short rod sections bent at right angles) imbedded in the hull membrane. They can either be fabricated before the hull is plastered, or afterwards, but starter rods must be placed before plastering.

Where pipe-frames or wooden frames are being used for the mould, webs will have to be fabricated to take the bulkheads. Metal flats or flanges can be welded to the reinforcement in their place to which bulkheads, etc. can be bolted, but webs are much better. They will give added rigidity to the hull, while the flanges could be pulled away during the flexing of the vessel and it would be extremely difficult to replace them. To fabricate the webs starter rods are poked through the inner layers of mesh in the plotted position of the bulkhead. Note that the webs for the floors can similarly be fabricated. In their case horizontal rods will run across the width of the keel and up to the desired height, joining the port and starboard hull sections. The ends of the reinforcing rod are bent at right angles, giving at least a 9 in. foot, which is wired firmly to the stringers. To make this job easier, it is an idea to place the rod starters before the outer layers of mesh go on. Otherwise time is spent in working the rod through the mesh from inside. Verticals will be needed for the floor webs to take the stress of forces acting directly down on the member. These are either wired in place or welded. Sufficient layers of mesh, say three on each side, can be applied and wired tautly to the 'grid'. The same construction can be carried out for the tanks, engine box, etc. Often the fabrication of these internal structures is left until the hull has been plastered and the mortar cured. More time can be devoted to finishing the hull and the fixing will not get in the way during plastering.

If the deck is to be in ferrocement, it too can be fabricated at this time, but in practice it is better left until after the hull has been plastered. But starter rods (extensions of the transversals) must be left bare so that there is some structure to bond to. On the outside of the hull the fittings for the rudder must be seen to. If they weren't already attached to the sternpost, rudder hangings (gudgeons) can be welded or bolted in place and a shoe or bearing fitted for the toe of the rudder post to rest in.

All skin fittings, exhaust, heads outlets and inlets, galley drain, log, echo-sounder, etc. can be located before the hull is plastered. Many builders advocate that they be placed at this time, warning of the difficulties of drilling through the hardened mortar, cutting the mesh, or hitting any of the rebar. However, from our own experiences they make too much of this problem. A good masonry drill, chisel, hammer and a hacksaw blade will allow you to place the through fitting without too much trouble. Perhaps the larger skin fittings can be located at this time, but it

31 The 39 ft ferrocement hull built by Manchester Ship Canal pilot John Law to a design by Samson Marine Design Enterprises of Canada. Before: the hull is wired up with all the reinforcement attached and is ready for plastering. After: the finished hull after a hard day's plastering, and the well satisfied labour force take a rest to break open the beer. At the time Law was planning on entering the yacht in the 1972 Singlehanded Trans-Atlantic Race. Photos: A. L. Edge.

is not necessary for most of them. If they are to be situated at this time, cut away the mesh to the desired diameter and put in a wooden plug or dowel. Let the plug protrude slightly on each side of the hull. I know of one builder who put in dowels which lay flush with the surfaces of the membrane. During plastering they were covered with a thin skim of mortar. I believe he is still looking for some of them! Another reason for not locating skin openings at this time (and this includes bolt holes to take chainplates, rubbing strakes, etc.) is that the dowels, sticking out as they will, obstruct the plasterer attempting to put a smooth finish to the hull. He would have to work round them, which will interrupt the 'flow'. We have also observed when the dowels were later removed, that there was a slight stippling (tiny voids) round the edge of the hole.

Hull ports and deadlights, unlike the other skin openings mentioned above, must be located now before any plastering takes place. Mesh and rod must be cut away to the right size and a plate steel flange shaped and welded in place. Or the opening can merely be blanked off with a piece of plywood, the fitting to be properly fixed later.

Again, before pouring in the mortar, check that the hull is adequately supported and is not likely to sag, or collapse in any way during the mortar application. In addition to supporting the hull from above on the rafters or scaffold, a few props are necessary underneath the structure. But during plastering they will cover areas which the plasterer will not be able to get at. When moved, after the mortar has cured, these untrowelled areas will show up clearly as uneven hollows in the smooth expanse of the hull. These can be plastered up, but it is never very satisfactory. An amateur builder here in England, Chris Cole, who at the time of writing is fitting out a 63 ft yacht, came up with a solution. To each frame at the turn of the bilge spot weld a short section of mild steel rod (a bolt would also be suitable) which will protrude from the hull. Now, attach to a strip of 2×4 in. timber a thin pipe section with a diameter just large enough to take the rod. This can fit over the gusset. This arrangement, providing it is properly fabricated, should be more than strong enough to support the hull (with a prop for each frame) and will not take up the area the conventional prop would occupy. Once the mortar has cured it would be a relatively simple matter to fracture the spot-weld holding each gusset, pull out the piece of rod and plaster the small hole left in the hull. This system has worked extremely well on at least two vessels built here in Britain.

One other point before plastering. The time lapse between the tying of the hull membrane to the big day when the mortar is poured is likely to be a few weeks. In that time, particularly if the hull is exposed, some rusting

32 The interior of the 65 ft yacht *Petralis* on plastering day. Mortar is supplied to the inside and forced through the mesh reinforcement to be plastered smooth on the outside. Note the prepared webs for bulkhead attachment and fabricated floor gussets. Scaffolding is supported on a pipe framework to allow the labourers to move through the length of the hull without standing on the hull itself, except inside the keel.

is bound to take place. This presents no problem, but the reinforcement should be brushed down well to remove any rust scale and dust which may have accumulated. This matter could impair the bond between the steel and mortar. But a certain amount of corrosion will pock the steel and help the adhesion of mortar. Any oil on the reinforcement must be cleaned off completely, preferably with detergents or degreasing solvent.
Paraffin is unacceptable, being itself oily.

ferrocement 9

MORTAR

Cement
Aggregates
Water
Admixtures
Mix design and preparation

The use of some form of mortar or concrete to bind together stone, gravel and other material for structural purposes has occurred from very early times. The Egyptians, for example, used a type of mortar manufactured from burnt lime to hold together the blocks forming the pyramids. The Etruscans and Greeks also employed a similar bonding material for the construction of their many edifices. The Romans, however, are the first people to have given any written description of the practice. One of the most notable examples of the application of mortar is the Pantheon in Rome. Its large dome, spanning some 142 ft, is constructed wholely in concrete. The concrete used was composed of limestone and a pozzolanic material which were burnt in a kiln. Mixed with water and an aggregate the composite turned hard. The pozzolana was a substance, still used today, which is found naturally in volcanic ash. Its name derives from the town of Puzzuoli in Italy near which the material was first collected. The pozzolana reacted with the lime in the mortar to make the material more durable.

Modern cement dates back to the mid-19th century. Its manufacturing process is different from that of the earlier cements and hardening occurs in different ways. It is often described as a hydraulic cement. You have often heard of the cement referred to as Portland cement. This name comes from a Leeds bricklayer, Joseph Aspdin, credited with being the first person to develop hydraulic cement. This he did by heating finely divided clay, limestone or chalk in a furnace hot enough to drive off the carbon dioxide in the material. The resultant product he named Portland cement. The prototype of modern Portland cement, however, was invented in 1845 by one Isaac C. Johnson, also an Englishman. Using limestone he increased the kiln temperature at which it was burnt to form a clinker, then pulverized it. The essential difference between modern hydraulic cement and that used by the ancients is that the hardening process, in the latter case, was more the result of the material drying out (hydration) and reacting with the carbon dioxide in the air. The hardening process of hydraulic cement on the other hand is due to a chemical reaction between the cement and water.

Concrete, or mortar, can today be described as a structural material formed by the chemical reaction between cement and water. This forms a hardened matrix which adheres to a mineral aggregate, which is usually sand. The properties of the mortar are governed by the type and quality of the constituent materials, the proportions in which they are combined, and the conditions. Following from this we can assume that if a mortar is to be suitable for a particular purpose one must select the constituent materials and combine them in such a way as to develop the qualities desired. In ferrocement boatbuilding we require mortar which will be strong, durable, weather resistant, impermeable and resistant to chemical attack from sea salts. It must be capable of providing a reasonably good finish.

Selection of materials and combining them to provide these qualities is not easy. There are a great many variables which can have an effect on the end product. The type of cement used can affect the strength of the mortar. Certain cements will provide strong mortars; others will not. Certain aggregates will produce mortar of greater compressive and tensile strengths than others. Another factor which can affect the strength of the mortar is the water/cement ratio. Strength will increase as this drops. Under normal conditions concrete will increase in strength with age, the rate of increase dependent on the type of cement used. Certain cements, for example, will produce a rapid increase of strength in a short period of time with the rate of increase after that period slowing down. A loss of strength can result if there is premature drying out of the mortar, and could be as much as 40 per cent. Temperature changes can affect the rate of set of the mortar. A rise in temperature will increase the hardening rate; a drop in temperature will slow it down.

Concrete is not totally impervious to moisture, but the degree of porosity can to some extent be controlled. In ferrocement the mortar can be made almost totally impermeable. Reduction of the water/cement ratio will result in an increased impermeability. A dense mortar containing a high percentage of cement to aggregate, with a fine aggregate, will be more impervious to moisture than will a mortar which is not as rich, or in which the aggregate is coarse.

So, there are numerous factors which have a bearing on the qualities of the mortar. They are related to mix design, the proper selection of materials, and the way in which the composite is handled and cured.

Cement

There are numerous types of cement on the market, but not all are suitable for ferrocement boatbuilding. There are five types which can be used and are recommended to the amateur or professional builder. All other varieties must be regarded with doubt.

Ordinary Portland cement (*BS* 12)

As its label indicates this is a common type of cement suitable for applications where special conditions do not prevail or particular properties are not required. It is a general purpose cement used for the construction of bridges, pavements, reservoirs, reinforced concrete buildings, etc. It should not be used for structures which are likely to be subjected to strong action from sulphates. This cement is often used in ferrocement boatbuilding in hotter climates as it does not generate as much heat as some of the other types of cements during hydration. In the US this type of cement is known as Type 1, with an ASTM C150 standard specification. In New Zealand the cement is named as in Britain. Its national specification there is NZS 1844.

Rapid hardening Portland cement (*BS* 12)

In the US this cement would be described as Type 111, national specification ASTM C150. In New Zealand it would conform to specification NZS 1844. This type of cement is selected when high early strength is required. Its strength gain accelerates rapidly in the first few days after plastering. The faster setting rate of this cement provides an advantage where the weather may be cold in that it will simplify the curing process. In such conditions it is often hard to maintain the right temperature gradient over a long period. In conditions where the temperature is low the hydration and strength development of the mortar proceeds more slowly than normal, and the curing period has to be extended. This rapid hardening cement might create problems in a hot climate: the setting rate might be too rapid to carry out proper finishing to the hull.

Sulphate resisting cement (*BS* 4027)
The US national specification for this cement is again ASTM C150. It is often referred to by North American builders as Type 5 cement. It is the cement most recommended for use in ferrocement boatbuilding, although it is not always readily available in some areas of the world. It has an average setting rate and will therefore not press the amateur builder into hurrying the job during plastering, which will probably result in a poorly finished hull. Sulphate resisting cement has a much better performance than ordinary or rapid hardening Portland cement. As its name indicates it will resist sulphate attack and will stand up better to prolonged exposure to seawater, or to hard fresh water. It is more expensive than the other two types of cements. The sulphate content of seawater is sufficient to affect non-resistant cements, but the effects are not so great as to make sulphate resistance the over-riding consideration in selecting a cement.

High alumina cement (*BS* 915)
High alumina cement is not a Portland cement. It is manufactured by fusing together in a furnace limestone and bauxite (aluminium ore). The cement consequently has a high proportion of aluminate, which may often be in the region of 35 to 44 per cent. High alumina cement has sometimes, but not often, been used in the construction of ferrocement vessels. I would, however, hesitate in recommending it to the home builder who might perhaps be having his first taste of this construction technique, for the cement is not as easy to work with as the Portland cements. The high proportion of aluminate results in a very rapid development of strength. Setting is accompanied by the generation of heat, so it is not considered particularly suitable for use in hot climates. It certainly could provide advantages in cold conditions. Because of the large evolution of heat during the chemical reaction between it and the mixed water, care must be taken when working with it. A further disadvantage of high alumina cement is that its compressive strength will fall when exposed to temperatures above 80°F, particularly where there is high humidity. Its main advantage is its high resistance to attack by sulphates. A warning: if high alumina is to be used, it must not be used in conjunction with ordinary Portland cements. The two do not combine at all well.

Portland blast-furnace cement (*BS* 146)
This cement is often recommended for ferrocement marine structures, but it is not often used as it is not readily available. It is mainly manufactured in Scotland. Cement clinker is combined with blast furnace slag in a ratio

of 35/65. The resultant properties of the cement depend very much on the character of the slag. Generally it has a slower setting rate than ordinary Portland cements, and a slight reduction in strength. The cement is also affected by low temperatures, so it is not recommended for the colder climates. Blast furnace cement is often used in concrete structures which have a large mass, mainly because there is a lower evolution of heat during hydration. It has particular advantages for the ferrocement boatbuilder. Its slower setting rate will give him more leeway to work the mortar into the steel reinforcement and finish the outside. Also, the cement is much more resistant to chemical attack, especially in seawater, than ordinary Portland cement.

Pozzolanic cement

One has often read references to pozzolanic cement being used in the construction of ferrocement craft. It is a material which is often used in Italy, but is not readily available elsewhere. Pozzolanic cement has similar properties to sulphate resisting cement, which is widely recommended by the ferrocement experts. It is resistant to chemical attack and has a medium setting rate, but has a somewhat slower strength gain than ordinary Portland cement. The cement is produced by grinding together 60 to 80 per cent cement and 40 to 20 per cent pozzolana. This latter could be volcanic ash, pumice, or an artificial product such as burnt clay or shale, and is substituted for up to 10 per cent of the cement in the mortar.

Your choice of cement must be made with respect to the conditions prevailing during plastering and curing. For most cases the sulphate resisting cement (Type 5 in the US) is recommended. It has a slower setting rate than most of the other cements. As most home builders will be hard pressed on plastering day, with an insufficient labour force to help out, this feature will facilitate the finishing of the hull. Then this cement is very resistant to attack by sulphates and other chemicals present in seawater. Where weather conditions are likely to be cool, which could affect the eventual setting time of the mortar, rapid hardening cement would be a help. But bear in mind that accompanying the high early strength characteristics of this cement is a reduced resistance to chemical attack. While high alumina cement has been used successfully in ferrocement constructions it could present problems during application. Amateur builders not fully conversant with concrete methods should avoid using it.

Bags of cement vary in weight according to country (50 kg in UK). Estimating the exact quantity one is going to need for the job requires some experience so it would be well worth your while asking advice from the cement suppliers or an engineer who is well versed in these matters. But bearing in mind that the density of the ferrocement should be close to 151 lbs per cu. ft, of which at least 120 lbs per cu. ft constitutes the mortar, a rough estimate can be made. Cement to aggregate is usually mixed in proportions of one part of cement to one-and-a-half to two parts of aggregate, by volume. Taken with the water content the top limit of cement would be about 25 to 30 per cent of the total volume of mortar. From there it is simple enough to work out how many cubic feet of cement will be needed. One 50 kg bag of dry cement would be about $1\frac{1}{4}$ cu. ft.

As any moisture absorbed will be detrimental to the quality of cement, only fresh cement which has not had the chance to pick up any dampness from the atmosphere should be used. It is not good practice to store cement too long before it is used, as no matter how much care you take, humidity will eventually air-set the cement, making it virtually useless. It is easy to tell whether a bag has gone bad or not from the presence of lumps in the powder, easily felt in the corners of bags. No opened or damaged bags can be used. If bad cement is used the mortar develops no strength and will crumble. So, take delivery of your cement only shortly before you are due to plaster the hull. Note that sulphate resistant cement (Type 5) is not usually kept in stock by suppliers. Place your order well in advance so that they can get a fresh supply in for you.

When the cement finally arrives on site care must be taken in storing it until such time as it is used. It is imperative that the bags be protected from all moisture. Storing in a shed which is completely dry, with all possible openings sealed, would obviously be the answer. If no such cover is available pile the load of cement under some waterproofing to protect it from rain or general humidity. Canvas tarpaulins or plastic sheeting are often used. Also, it is a good idea to raise the bags off the ground on a wooden platform or some similar structure.

Aggregates

Aggregate is the term given to the mineral, usually sand, which is mixed with the cement and water to make up the mortar. The quality and type of aggregate used are important for the production of a high quality mortar which, for ferrocement structures, must be strong, impermeable, sufficiently workable to achieve proper penetration of the mesh, and capable of providing the smooth finish necessary on a boat hull. There are many different types of aggregate which have been used in reinforced concrete but not all are suitable for ferrocement, so it is important that the right material is obtained.

Aggregates used in reinforced concrete structures are loosely grouped under two headings: coarse aggregates and fine aggregates. For our purposes only the latter need be considered. Fine aggregate is that which will mainly pass through a $\frac{3}{16}$ in. (4.76 mm) sieve complying with BS 410 (or the equivalent national standard in other countries). This aggregate is normally natural sand which has resulted from the disintegration of rock, or crushed stone sand, or crushed gravel sand. These aggregates produce the best type of mortar. Silica sands are often recommended for ferrocement, but they are not always conveniently available. Both particle size and shape affect quality.

Any aggregate used must be hard, durable and free from any impurities which could affect the bonding properties of the mortar. Such materials include clay, silt, fine dust and other adherent coatings which will weaken the cohesion between the particles that make up the mortar. Such coatings can often be eliminated by careful washing of the aggregate. No organic substances like coal or sulphur, or soluble salts, which may include calcium, sodium and magnesium, should be present in the aggregate. They could cause swelling or disruption of the mortar through chemical reactions, or corrode the steel reinforcement.

Other materials which are not desirable in an aggregate are those in which the particles are flaky or elongated. The best possible shape for the particles which constitute the aggregate is round or cube shaped, for this will allow the material to be compacted far more satisfactorily and make for a more workable mix. There is less possibility of minute voids occurring as cement will more easily distribute round them. The result is a much denser and stronger mortar. Laminated materials such as mica and shale should not be present in the aggregate, for they will weaken the cohesion of the mortar.

Reference is sometimes made to 'lightweight' aggregates, some of which have been employed in ferrocement boatbuilding. These aggregates may include clinker, foamed slag, expanded clay, shale or slate, vermiculite and pumice. Their use in place of natural sands is an attempt to reduce the weight of the mortar, an obvious advantage in boat construction. Expanded clays, for example, are said to give a weight reduction somewhere in the order of 20 per cent. During the last war this type of lightweight aggregate was successfully used in the construction of large reinforced concrete ships. I know too of a yacht which has been built using pumice as an aggregate. But such materials must be treated with suspicion (shale, slate and vermiculite are in fact laminated), and are not generally recommended for ferrocement hulls, for there is a loss of strength and increase of porosity accompanying their use.

The size of the particles in the aggregate and their distribution through the mass is important. As I mentioned earlier only fine aggregates which will pass through a $\frac{3}{16}$ in. sieve should be used. But then the material must further be graded into fines and coarse particles within this gauge, and covering a range which will produce a dense, strong and impermeable mortar. The grading of fine aggregate is achieved by sieve analysis, using sieves complying with BS 410. The method to be followed is specified in BS 812, 'methods for sampling and testing mineral aggregates, sands and fillers'. Graders Limits are in BS 882 and 1201.

The sieves used for the grading of aggregate are sized according to the number of opening per square inch. The number designates the quantity of openings. A grading chart is included to indicate the size distribution necessary in aggregate used for ferrocement. Experience has determined that these aggregate grading limits will provide mortar exhibiting the most satisfactory results. A deviation to a higher percentage of coarser particles will impair the workability of the mortar, reduce its strength and increase the difficulty of achieving good penetration and finish. Deviating to the other extreme, using an aggregate with too high a percentage of fines will increase the demand for water. This will in turn be followed by a drop in strength, greater porosity, and shrinkage. Reading from the chart you will see that 100 per cent of the aggregate must pass through a No. 7 sieve; 68 to 95 per cent through a No. 14 sieve; 35 to 65 through a No. 25 sieve; 10 to 35 per cent through a No. 52 sieve; 2 to 10 per cent through a No. 100 sieve; and none through a No. 200 sieve. The sieve equivalents for the United States are included on the chart.

If possible it is always best to obtain your aggregate from a reputable supplier. When an order is placed BS 812 should be quoted. The sand you obtain can then reasonably be expected to have been graded and washed to comply with that specification. But it might be necessary for you to again sieve the load in order to comply with the grading limits. It is always a good idea to send samples of the aggregate away for testing. Naturally if your source of supply is from somewhere other than a commercial supplier, you will need to have samples tested to check the suitability of the material.

Even if you have obtained aggregate complying with the necessary standard it might often happen that the material has become contaminated in some way before it is actually used. This can easily happen if the sand

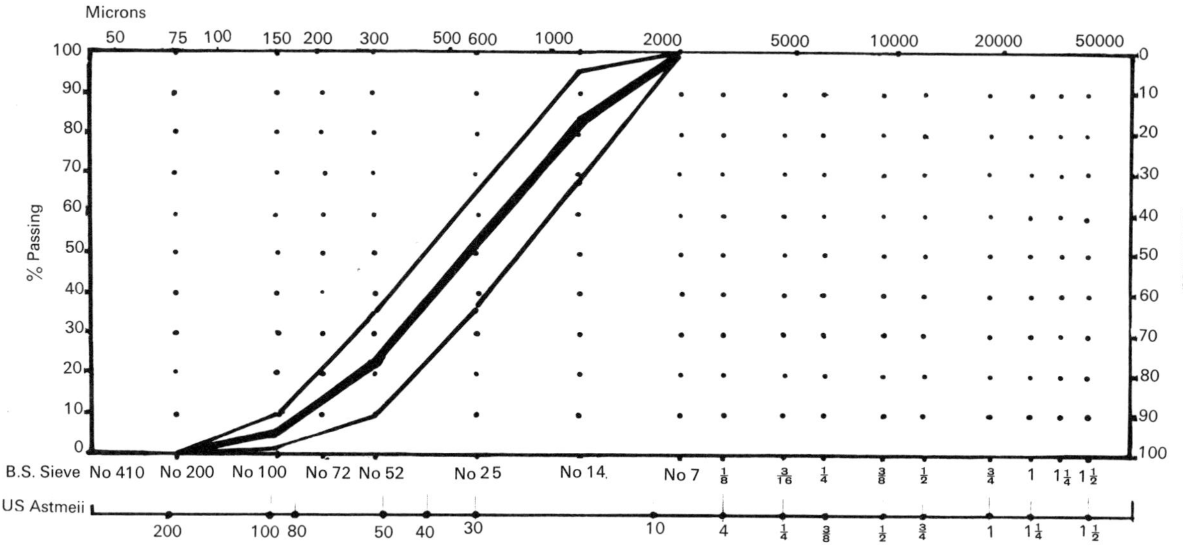

Grading chart indicating a suggested envelope for ferrocement mortar sand, reproduced with permission of the Cement Marketing Co.

has been on site for some time, during which it could well have collected a load of rubbish. In such an instance a few simple tests can be carried out to reaffirm the quality of the material. It might be that you will have to wash the sand and screen it again.

Bad contamination can normally be observed directly. Sticks, leaves, lumps of clay or rock and other extraneous matter which the sand may have collected can be screened out. Excess fine material such as fine clay or silt can be indicated by a simple sediment method. Put about 3 in. of the sand in a glass container—a milk bottle or one of those large preserve jars would do the trick. Then fill the jar with water to about an inch or slightly more above the level of the sand. Shake the bottle and leave the saturated sand to settle. The coarser particles will drop almost immediately but the finer material will remain in suspension and settle more slowly, eventually forming a visible layer on top of the sand. The depth of this layer should not exceed 5 per cent of the total depth of sand. If common table salt is added to the water at the beginning—a teaspoonful will be sufficient—settlement of the finer particles will be accelerated. Using plain water it would be necessary to allow the settling to take place over 24 hours. With salt added it could be as little as three to four hours.

The presence of any organic impurities in an aggregate are not immediately observable, but there is a test which can be carried out on site to determine roughly what contamination there may be. Understand that any such materials could retard the hydration of the cement, and impair its adhesive qualities. In this instance obtain a representative sample of the sand, again placing about 3 in. in a glass jar. In place of the water use a 3 per cent (wt/vol) solution of caustic soda. Presence of any organic compound will be indicated by browning of the solution as the compound reacts with the caustic soda. A dark brown colour will indicate serious contamination; a lighter colour will indicate little contamination.

Water

The most important consideration concerning water used for mixing the mortar in ferrocement, or any concrete for that matter, is that it should not contain any substance likely to influence the setting rate or quality of the mortar. Such substances can include clay, loam, acids, soluble salts, decaying vegetable matter and any other organic substances. Such deleterious materials are more than likely to be found in rivers, dams and perhaps boreholes. If these sources of supply are being considered, get samples laboratory tested first. But as a preliminary you could possibly check the water by comparing samples of mortar, where one sample is made with distilled water and the other with the suspect fluid. If the latter sample in any way fails to meet the qualities of the former, you may be sure that the water has something to do with it. In such a case you could have further tests made to see whether some remedy is available for clearing the water. This step need only be resorted to, of course, if this is your only convenient supply of water. Seawater should not on any account be used for mixing the mortar as it tends to make concrete hydroscopic. It could be used as a last resort for wetting down the hull during curing, but I stress only as a last resort. Usually water that is piped from public supplies is regarded as satisfactory and does not require any further testing. As a last word, bear in mind that you are going to need a fair amount of water, both for the mixing of the mortar and lastly for curing the hull. Make certain that you have a sufficient quantity on tap to meet your requirements.

Admixtures

Admixtures ranging from chemical additives to a variety of waste materials introduced to the mortar to alter or improve one or more of its properties have been advocated from the time when cement was first invented. There are numerous chemicals and other materials which can be added to mortar with the intention of altering one or more of its properties. But it is important that any admixture used is fully vetted before being introduced to the mix. It must be emphasised that although certain benefits can be derived from the correct use of such additives there is no substance available which will compensate for inferior materials or bad workmanship. Concrete of poor quality will not be improved by any additive.

Admixtures can be classified into groups, according to the effect they are supposed to accomplish. At present there must be about 300 or more different types of admixtures available in Britain, with this quantity increasing all the time. Chemical admixtures containing calcium chloride should not be included in any mortar used for ferrocement, for they increase the risk of corrosion of the reinforcements.
Therefore when selecting an additive or admixture for a particular purpose ascertain first that it does not contain any calcium chloride or nitrates.

The main reasons for using admixtures in concrete is to improve workability, durability or to acclerate or retard setting and hardening. Such admixtures are only introduced in minute quantities to produce the desired effect; therefore the degree of control must be high. Often the trouble and expense involved in achieving this control can far outweigh the advantages derived from the additive. Excess amounts of the material, whatever it may be, can be highly detrimental to the strength and other properties of the mortar. Additives should be avoided unless their reaction with the cement is properly investigated and considered satisfactory.

Admixtures can be classified as water-reducing agents, retarding agents and accelerating agents. The first is designed mainly to increase the workability of the mortar without the need to step up the water content. Or the water/cement ratio can be decreased and by adding the necessary amount of agent the workability can remain constant. This will also allow a strength gain, increased impermeability and increased durability. The retarding admixtures will slow down the rate of set of the mortar, which could be beneficial in warm weather where problems may arise in the finishing off of the hull. Ultimate strength of the mortar may be slightly higher than for undosed mortar, but since strength development is retarded during the early stages of set, it will almost certainly be necessary to continue the curing process for a couple of extra days to obtain full benefit. Extreme care must be taken that an overdose is not given, for it can seriously affect the setting of the mortar. In any event this type of admixture is seldom used by ferrocement builders. The accelerating admixtures, the third in the group, will do as their label indicates: speed up the setting rate of the mortar, or the early strength development of the mortar. Their use would generally be limited to cold environmental conditions where lower temperatures could impede the progress of the set. However, should such conditions prevail it would be far better to make use of rapid hardening cement to achieve the same results. One should be very careful with accelerators for most of them are based on chlorides or calcium chloride, which must not be present. Besides these three main groups there are mixtures or reagents which will demonstrate the characteristics of a combination of two of them, e.g. water-reducing and retarding admixtures, and water-reducing and accelerating admixtures.

Of all these agents the most commonly used or recommended are the water-reducing admixtures. I cannot emphasize too much the importance of following to the letter the manufacturer's instructions. And on no account should excessive amounts be used.

A water-reducing admixture which has been successfully used here in Britain on a couple of boatbuilding projects is a Blue Circle product given the name Lanomix. It has been found to be most useful when a mortar pump has been employed to apply the mortar to the hull. It gave the mortar that extra bit of workability necessary to force it properly through the pump. According to the blurb Lanomix will reduce bleeding, segregation and laitence and increase workability. It aids compacting and by reducing the water/cement ratio without loss of workability will provide higher strength and reduced shrinkage. Only small quantities must be introduced to the mix, and that before mixing takes place. It is supplied in liquid form and, depending upon the properties required, instructions specify that only $\frac{1}{4}$ to $\frac{1}{2}$ pint be added to 50 kg of the cement. The inclusion of Lanomix into the mortar is likely to result in slightly slower setting times and slightly increased air content (about 1%).

This brings up the subject of plasticizers or air-entraining agents which are sometimes mentioned for improving the workability of the mortar. The chemicals in them introduce tiny air bubbles into the mix making it more 'fatty' and easy to work, and reduces the segregation of cement and aggregate. Such admixtures may also be described as water-reducing agents. They must be used with extreme caution. Their use will introduce air bubbles, sometimes to the extent of 1 to 5 per cent by volume. Accompanying increased workability is the very real danger of strength loss of the mortar which could be as much as 20 per cent. Porosity may also increase. In general many ferrocement technologists hesitate in recommending such admixtures.

Additives described as pozzolans are perhaps the most common materials which builders advocate for inclusion in the mortar. They are added with the purpose of improving workability and durability; for increasing density; increasing impermeability, and thus reducing the dangers of damage occurring through chemical attack. They will generally improve the sulphate resistance of the mortar. The pozzolans contain constituents which will combine with the free lime present in the mortar to form insoluble compounds which possess cementing qualities. The result would be to increase or augment the strength of the mortar.

Pozzolans may include fly ash, diatomaceous earth, bentomite, kaolin or blast furnace slag. When properly ground all have a high percentage of fines which will aid the density of the mortar, acting as pore fillers. Again these additives must be used with caution. Unless their reaction with the cement is established and considered satisfactory they should not be used. Fly ash, for example, may often contain impurities in the form of unburnt carbon or sulphur which, as previously commented on, will have a deleterious effect on the cement. Suitability of different pozzolans should be tested and verified. Sometimes useful information can be obtained by comparing trial mixes with and without pozzolan added. Since pozzolan reacts with the free lime in the mortar which has been released by hydration, the full benefit of its inclusion will not be immediately realised

In general it is recommended that not more than about a 10 per cent substitution of pozzolan be incorporated into a mix, the material to replace part of the cement and not the aggregate. Some builders have gone as high as 15 per cent. Samson Marine Design Enterprises suggest an approximate ratio of around 5 lbs of pozzolan to a hundredweight of cement. Other builders fall in line with this value. However, many professional ferrocement boatbuilders in Britain, the US and New Zealand have largely discontinued using any pozzolan in their mixes. Instead they have concentrated on obtaining a good properly graded aggregate, preferably with a high silica content. The exclusion of pozzolans from the mix has thus far not indicated any significant loss in quality of the mortar. In cases where a sufficiently good quality aggregate is not obtainable there might be some grounds for introducing a pozzolan to the mix.

Mix design and preparation

The mix design of mortar used for ferrocement boatbuilding is probably more critical than for many other concrete structures. While builders have been known to differ slightly over details of design, generally they confine themselves to fairly narrow limits, using a high concentration of cement and a relatively low water content, just sufficient to maintain necessary workability of the mortar.

The optimum mix proportions will depend to some extent on the materials used (different sand gradings, for example, may alter slightly the mix proportion) and can only be properly determined by preparing a few trial mixes. A high cement content will normally make for a strong, impermeable mortar capable of giving a reasonably fine surface finish. Reducing the cement content will bring about a drop in strength. In practice the composition of mortar used by ferrocement technologists varies between 50 and 60 lbs of cement per cu. ft of sand. On no account should the proportion of cement drop below the lower limit. In different terms, the usual cement/aggregate ratio recommended is in the region of 0.56, which will be in the order of one part of cement to 1½ to 2 parts of sand, measured either by weight or volume. A much higher proportion of cement may be used, up to about 0.66, but accompanying it will be an increase in shrinkage. Above a cement/sand ratio of 0.66 there will be a drop in strength of the mortar.

MORTAR MIX PROPORTIONS (based on whole and half cwt bags cement)

Cement	(lbs)	(lbs) *Dry Sand*	(cu. ft)	(gals) *Water*	(pints)
½ bag	47	70–94	¾–1	1	5
	56	84–112	1–1¼	2	–
1 bag	94	140–188	1½–2	3	2
	112	168–224	1¾–2½	3	7
1½ bags	141	222–282	2¼–3	5	–
	168	252–336	2¾–3½	5	7
2 bags	188	282–376	3–4	6	5
	224	336–448	3½–4¾	7	7

(Table prepared by New Zealand Portland Cement Association. Slight adjustment of the figure is necessary for metric or other bag sizes.

The water/cement ratio must be kept as low as practicable. Water has a considerable effect on the strength characteristics of a mortar. In simple terms the strength of the mortar will decrease as the water content increases. Water has two functions in a mix. In the first place it is needed to set off the chemical reaction of the cement and bring about hydration. Secondly, it lubricates the mixture to facilitate mixing. Only very little water is needed for hydration. With normal Portland cement, for example, one part by weight of the cement will require only about 0.25 parts by weight of water for hydration. Such a mix would be far too dry and difficult to work, so extra water is necessary to improve the workability of the mortar. In ferrocement the water/cement ratio is regulated to the region of 0.35. This would work out to roughly 4 gallons of water per hundredweight of cement. A table is included which can be used as a guide to indicate approximate proportions of cement/sand/water. Cement is mostly purchased in bags weighing 94 lbs or 112 lbs so the table is worked on the basis of whole and half bags. The figures were supplied by the New Zealand Portland Cement Association.

The best and most accurate means of preparing batches of cement and aggregate is by weight. A scale of some sort will be needed. However, where a scale is not available batching can equally be carried out by volume measurements. If care is taken one can obtain reasonably consistent batches. Most builders in fact do prepare their mortar by volumetric means. Whether batching by weight or volume it makes practical sense to determine the mix proportions on the fact that cement is supplied in bags of constant weight. Where measuring by volume suitable measuring containers must be provided. Large sized buckets, measured and marked with desired levels, can be used. Sometimes a standard box with dimensions providing a capacity of 1 cu. ft is constructed to assist the job of measurement. Other containers, cans, or tins of different sizes may be handy for fractional measurements. A gallon can will, for example, equal about a sixth of a cubic foot.

Water is usually measured in gallons and pints, so there is no problem in providing the necessary measuring containers. When determining the right quantity of water needed for a mix, the moisture content of the aggregate must be taken into account. Most sand contains a certain amount of water, absorbed mainly from the atmosphere. This could be in the region of 4 to 8 per cent. Of course it can be far more if the sand has been saturated by rain. All the mix proportions referred to above and in the table relate to dry sand only, so certain adjustments have to be made during batching, of both sand and water, to achieve the right balance.

When the aggregate is partially wet it will bulk, or increase in volume through swelling. One cubic foot of partially wet sand may, for instance, really contain only about three quarters of that volume of dry sand, the rest being made up of two or three pints of water. This will prove to be the case more with fine than coarse aggregates. This characteristic has particular relevance if one is going to batch the mortar ingredients by volume. There is a simple method for determining the approximate amounts of bulking, keeping in mind that sand which has been wholely saturated will occupy the same volume as dry sand. Half fill a glass bottle or beaker with a representative sample of the aggregate. Take the height measurement, and then saturate the sample with water, tamping it to ensure that all air bubbles are removed. If the sample had previously contained a certain proportion of water you will note that the level will have dropped slightly. Measure the difference and by simple proportion the percentage bulking can be worked out.

There are a number of standard methods of determining the moisture content of an aggregate. The methods are described in BS 812, but most of them require laboratory facilities. One method can be attempted in the home. This involves weighing a small sample of the aggregate and then drying it at a high temperature in an oven, and weighing it again afterwards. The possible drop in weight can be termed as a percentage of the original weight.

In practice many experienced builders and plasterers go by the 'feel' and sound of the mortar to determine whether or not they have achieved the right water content and workability of the mortar. This is done by adding only a portion of the necessary water to the mix; allowing this to be worked in; then adding the rest little by little until the mortar has reached what appears to be the right consistency. One standard test which is often used on site to determine the workability is known as the 'slump test'. This employs a cone made of metal and open top and bottom. A batched sample is dumped into the cone until it is completely filled. Tamp the mortar with a rod to compact it and skim the top flat. Now remove the cone. If the mortar is too moist it will have slumped slightly. For mortar used to plaster boats there should be as little slump as possible. Always keep in mind the necessity of keeping the water content as low as possible. If water must be added for the sake of workability do it sparingly. The idea is to wet the mix just enough to make it possible to work it into all the gaps in the mesh, without adding so much water that it sags or falls from the reinforcement.

The same care must be exercised when any admixture such as a water-reducing agent or a pozzolan is introduced to the mix. Go strictly according to the manufacturer's instruction and don't give the mix an overdose.

ferrocement 10

PLASTERING THE HULL

Labour force
Plastering day
Single-stage and two-stage methods

The plastering of the hull is often considered the most critical phase in the whole ferrocement boatbuilding process. And it could well be, as any mistakes or errors occurring now could spell out failure of the total operation for they will be most difficult to rectify. During the other stages of construction any mistakes occurring can to some extent be corrected. Mesh can be rewrapped; high tensile rod stringers and transversals can be stripped off and relocated. Even individual frames can be removed for replacement or reshaping. In fact the complete hull basket can be dismantled if need be, though this drastic step is not to be recommended. Once the mortar is mixed and partially or wholly applied to the hull there is little chance of being able to turn back if a fault is discovered. This situation affects many people psychologically, causing them to look to the approaching plastering day with some trepidation and tightening of the stomach.

However, if you make careful preparations well in advance of the day so that nothing, but nothing, is left to chance, you could avoid any likely pitfalls. It's always a good idea to draw up a check list and mark off each item as it is prepared for. First on your list should be the hull. Spend a couple of days going over it carefully ensuring that it is well and truely ready for plastering. Run a wooden batten over it to check that all the reinforcement conforms to the desired hull shape and is as fair and tight as it possibly can be. Any loose bits of wire and protruding tie ends should be clipped off flush with the mesh membrane or bent back into it. If these are left sticking out they would hinder the plasterer attempting to trowel smooth the surface of the hull. But most serious, they would eventually corrode, disfiguring the hull and leaving an open pore in the shell giving access to the imbedded reinforcement. Further corrosion could occur, the consequences of which are pretty obvious. These ends can be bent back during trowelling, but it is best to do the job before plastering actually begins.

Check that all the necessary through hull skin fittings have been placed. Only for large ports and other big diameter cavities is it better to prepare them before plastering. Look to your interior fabrications and make sure you have them all either in place or prepared for with starter rods. Other fittings like chain roller, chainplates, fairleads, etc, which may be incorporated in the hull should be positioned prior to plastering. A wooden or metal capping piece can be laid along the deck line, toerail or bulwark to give the plasterers a regular line to work up to, and make for a neater edge. It can either be tack welded or wired in place. The latter is preferred.

Most important is to check again that the hull is properly supported by props and the overhead support structure (if there is one) so that there is absolutely no likelihood of the hull moving during the application of the mortar, or deflection occurring. But first bear in mind that the underneath of the keel must be plastered. This can be done by suspending the framework from the cross bearers or rafters, removing the timber support from below the keel and slapping on the plaster. Take it about 3 in.–4 in. up the sides of the keel and attempt to get maximum penetration. Then return the timber support beneath the keel, set the hull firmly down on it and put in the side props firmly. This early mortar can later be keyed when the rest of the keel is plastered so that there is a good strong bonding at the cold joint. A preliminary cement grout or a coating of epoxy resin (the usual two-pot variety) will improve adhesion at the interface. And anyway there is little danger of any real leakage occuring for the joint will be well below the open trough level within the keel. Instead of plastering the base of the keel at this stage it can always be done later, when the rest of the hull has been completed and cured, but you will have the extra weight of the hull to support if you leave it until the end.

Particularly if the fabrication of the hull has taken place outdoors, exposed to the weather, there is quite likely to be a build-up of rust on the reinforcement. Normally it is only a very slight surface coating, but it is a good idea to take a wire brush to it, taking away any mill scale and excessive areas of rust. A certain amount of rusting will assist the adhesion of the mortar to the reinforcement. The main cause of bonding between the two is the shrinkage of the mortar as it cures. Added to this is the mechanical keying at the mortar/steel interface, which will be assisted by the slight pitting of the steel from rusting. To further aid bonding some builders suggest coating the reinforcement with a light layer of epoxy resin. In practice I feel that this is going a bit far, for it is not really necessary, though it won't do any harm. For the amateur, who is perhaps working on a tight budget, it is a needless expense. Epoxy resin is not a cheap commodity. Any excessive coatings of oil, grease and other matter which may harm the mortar and affect its bond to the steel should also be removed before any plastering is done. A dilute solution of caustic soda will be of use to wash down the hull armature. A much greater degree of cleaning and degreasing will be necessary if a coat of resin is to go on: for the mortar it is less critical.

One last item in connection with the hull. Make sure that there is easy access to every part of it, with no obstructions to impede the plasterers. Some sort of scaffolding will be needed both inside and outside the hull to allow the application of the mortar without any undue weight being placed on the reinforcement. An attempt must be made always to reduce the dangers of deflection. Workers trampling on the reinforcement could cause the mesh and rods to bend out of true, which may not be immediately noticeable. Bulges which occur will have to be covered up by an extra build-up of mortar, which is undesirable because of the unnecessary increase in weight and an increased tendency to surface micro-cracking. Micro-cracks will only travel as deep as the mesh membrane, where the interspersed steel fabric will absorb the tensile stresses, but attempts must be made to limit the micro-cracking on the surface and any excessive build-up of mortar is not going to help.

The normal arrangement adopted inside a hull is to lay a few planks lengthwise down the hull. They can possibly be supported by the trusses or frame spreaders connecting each pair of frames, which are used to strengthen and stiffen the framework. Or, in the case of prefabricated frames, the planks can be supported by the floor webs, suitably strengthened with timber beams. Some builders employ a series of angle iron or T-bar sections which are tack welded to each pair of frames. The scaffolding can be carried by them. Another idea is to drop a series of vertical supports from the overhead cross bearers or rafters, if there are any, and weld a horizontal to each of them to take the planks. Outside the hull a working platform can be set up right round the hull, preferably halfway up the side. Perhaps some means of making slight height adjustments would be beneficial if the hull has very high sides. Anyway, whatever scaffolding system you do employ, make sure it provides a rigid working platform. A rickety affair will only distract the plasterers from the important job at hand, as they will always be too concerned for their own welfare. And remember a considerable amount of force is necessary to apply the mortar properly.

Draw up a special list of materials and tools which will be needed on plastering day and get them on site. Start with the most important items first and work down. First you must get in your load of cement, aggregate, any admixtures you may require, and make sure that the water supply is not going to let you down on P-day. It might be an idea to have a $\frac{3}{16}$ in. sieve available just in case the sand becomes contaminated with bits of stick, leaves and other rubbish while en route to the site or while it is there. I know of a number of plastering days here during which the aggregate has had to be rescreened in a hurry. In ordering the cement take a gander at the prevailing conditions. Difficult to predict the weather I know, but you should be able to gauge within reasonable limits what the temperature gradient is likely to be. If you're to plaster during a very cold day it might be a good idea to use rapid hardening cement. Alternatively plug for the sulphate resisting cement which has a much slower setting time. Don't be a 'Scrooge' in ordering the cement and sand. They are not the most expensive items on your short list, and you could find yourself in hot water if you end up short of supplies halfway through plastering. It is always good practice to overestimate slightly what your demand is going to be. You can always put the excess to good use later when the time comes to plaster in your

33 Mortar being applied to the top of the 63 ft hull built by Chris Cole via a flexible pipe from a mortar pump. The hopper of the pump can be seen in the background.

ballast. A small supply of epoxy resin might come in handy on plastering day to coat exposed skin fittings, like rudder gudgeons and stemhead fittings. The epoxy will improve the bond between the mortar and the fittings, and also prevent them from rusting up.

Near the top of your check list you should have a concrete mixer marked down. Don't for one moment consider the possibility of mixing the mortar by hand. That might be all right for laying a concrete path in your back yard, but most definitely not when a hull has to be plastered. It is a slow, inefficient process. Mortar used on the hull must be mixed properly and to the right consistency, which only a motorized concrete mixer can accomplish.

There are several types of mixer which could suitably carry out the job. There is no need to buy one, unless you are aiming to turn professional. They are usually rented, the fee being not too expensive. Anyway you are only going to need them for a day or two at most. The standard type of mixer used by many amateur builders here in Britain is the ordinary tilting drum type, which can be seen on almost any construction site. With this type the drum is revolved by a diesel motor, tilted by a hand wheel worked by the operator. The mortar inside is mixed by being thrown round the drum walls. This type of mixer has certain drawbacks, and is not generally recommended by experienced ferrocementers. Mortar used for plastering must be of a fairly stiff nature and the partially mixed mortar has a tendency to cling to the sides of the drum and not mix completely. Lumps with dry centres composed of cement or sand may sometimes form with this type of mixer. The person operating the machine must keep a close watch on the happenings inside the drum, regularly scraping the sides clean to keep the ingredients mixing properly.

The most suitable, recommended mixer is that which has a paddle attached to the inside of the drum to keep the mortar mixing evenly. The mixer is often referred to as a 'plasterer's mixer'. The drum revolves horizontally with the dipped paddle either turning with it at a different speed, or in the opposite direction. The plasterer's mixer produces a mortar of a better consistency than the tilting type and, in practice, you will find that less water will be necessary to obtain suitable fattiness or workability of the mortar. For vessels up to 70 ft in length we have found that one large-sized mixer will be able to keep pace with demand. But it would be handy to have two on site on plastering day, if only as insurance against one of them breaking down.

Further equipment you will need for the mixing operation will include a couple of shovels, several large buckets, or other containers with which to measure out the quantities of cement and aggregate. A couple of gallon, quart and pint tins will come in handy for measuring part quantities of cement or aggregate, or an additive like pozzolan. If you're intending to use some sort of admixture it is good practice to measure out the necessary dosages for each mix and put them in separate packages or containers. This will expedite the mixing operation and eliminate possible confusion that might arise when plastering is in full swing. Other buckets, eight to a dozen, will be needed to transport the prepared mortar from the mixer to the hull.

A hose will be needed to pipe water directly to the mixing area. It would help if some sort of valve could be fitted to the end of the hose to control or shut off the water when it is not needed. This measure would prevent the area directly round the mixer from rapidly becoming flooded. A 44 gallon drum or other container of similar size would be useful to keep clean water in. It would be easier to measure the right amount of mixing water from this container, than to turn the hose on every time. Talking of being flooded out, some builders erect a mixing platform, somewhat elevated so that all extraneous water can drain away.

For the actual job of placing the mortar on the hull most of the tools used in normal plastering operations can be put to good purpose. Each plasterer should have a set. They include floats and steel trowels, a wooden float and a hawk, which is merely a board with a wooden handle attached to it and used to hold a lump of mortar. It is easily made out of any off-cut. Then a trowel sponge, or just a sponge, will be needed to finish off the hull, smoothing it down and taking off any trowel marks, and a couple of those brushes commonly used for whitewashing, with long softish bristles. These are soaked with water which is flicked onto the hull during the finishing operation. Two or three clean buckets would come in handy to contain water and cement grout, also used to give a smooth finish to the hull.

In addition to the usual tools, there could be a call for some specially shaped equipment to cater for the many compound curves one finds on a boat hull. A flexible batten, sometimes called a derby, will come in useful for the flatter curves to help achieve a good smooth finish and fairness. This batten can be made either in wood or using a flexible strip of steel. A good size is about 3 ft long, about $1\frac{1}{2}$ in.–2 in. wide by $\frac{1}{8}$ in.– $\frac{3}{16}$ in. thick. A handle at each end will be helpful for manipulating the batten. For the really sharp curves a tablespoon or similarly shaped object is often used to form the mortar and press it well into the mesh membrane. As an alternative to the whitewash brushes, or in addition to them, a long-handled soft broom can be used to brush away trowel markings when the mortar has reached a semi-crumbly stage. Each member of the labour force who may be handling the mortar would best be equipped with strong pair of gloves. Those thick leather or rubber gloves sometimes used by gardeners or welders will do the trick. The mortar can play havoc with your hands if they're continually exposed to it. The gloves provide protection against loose bits of wire and sharp edges which could take a piece out of you.

A vibrator is necessary during the plastering. This is a machine, either electrically operated or driven by a two-stroke motor, which is used to vibrate the implaced mortar, allowing any captured air bubbles within the mix to rise to the surface. The vibrator helps to consolidate the mortar, ensuring better penetration and lessening the chance of voids being left within the shell. The vibrator most used by boatbuilders is known as a pencil vibrator. It has a prod about 2 ft long attached to the end of a flexible hose. When operating the prod vibrates rapidly and can be held by hand and directed over different areas of the hull. This item of equipment can be hired at the same time as the concrete mixer.

Now, to all intents and purposes you have all the basic materials and equipment you are likely to need for plastering the hull. But a further word on application. Many builders have successfully made use of what is known as a mortar pump, which is a machine which aids the swift delivery of the mortar from the concrete mixer to the point of application on the hull. It is a luxury item which greatly facilitates the placement of the mortar and is to be recommended. The pumps are not expensive to hire and allow more time to be devoted by the labour force in tamping in the mortar to achieve good penetration. It therefore also gives the plasterers more time to work up a good finish on the outside of the hull before the mortar sets. The machine has a hopper into which the freshly mixed mortar is dropped. The mortar feeds through a flexible pipe which can be directed at will. The mortar is forced through by the action of a screw-type pump which is driven in most cases, depending on model and make, by a compressed air motor. One model of which we have had personal experience is the Power-Spray Pump (Mk VII), produced by a Middlesex firm. This particular model has a hopper capacity of 5 cu. ft. Its output is in the region of 1 cu. ft of concrete per minute. This size of mortar pump has been successfully used for plastering hulls up to 65 ft overall. In using the pump we found that it was necessary to have a mortar which was slightly more workable than usual. A water-reducing agent eliminated this problem.

I have described above the Gunite process in which the aggregate and cement are forced through a pipe by air pressure, mixed with water at the nozzle and then sprayed. It is a quick method of application and achieves

good penetration, but it is a technique which I would hesitate to recommend.
The method demands an increase in the water content, which is to be avoided if at all possible as this will reduce the strength of the mortar and increase the problems of shrinkage. I have had reports of a couple of ferrocement boats plastered in this way. In one, which was relatively successful, there was still a high degree of surface fracturing of the mortar which had to be filled in. In another, the boat was a failure. The degree of shrinkage was very high indeed and large cracks formed in the shell. The mortar itself did not develop much strength. Nevertheless the Gunite process has been used with some success, particularly in the US. Fibersteel Corp. use the technique, which they prefer to describe as 'velocity placement', for applying mortar in their cavity mould process.

With all the equipment and materials arranged for, consider next the logistics of the plastering operation. Select the site for the mixing operation and have all items needed for that job close at hand. Where use is to be made of admixtures, like water reducing agents or pozzolan, it is a good idea to have each measure prepared beforehand and placed conveniently close to the concrete mixer. Keep the cement bags and sand a little out of the way of the concrete mixer so that any waste water will not reach them. Don't select a mixing site too far from your hull. Buckets of mortar can be heavy, and they'll get heavier as the plastering progresses, so situate for ease and swiftness of delivery. A wheelbarrow would be handy to deliver the mortar to the far ends of the hull. As mortar will be placed within the hull, a couple of block and tackles will make life a bit easier.

Labour force

I think that one of the most pertinent reasons for the slow or reluctant acceptance of ferrocement among the yachting fraternity is the relatively poor finish that many of these concrete vessels have had. While being believably strong and seaworthy the boats have tended to look rough and amateurishly made alongside any other boat constructed say in plywood or glass reinforced plastics. The hulls have had ripples and bumps in them and the marks of the plasterer's trowel have been clearly visible. There is absolutely no need to produce a hull looking that poor. Plastering a hull may look relatively simple, but the touch of the professional will stand out a mile. So, if possible, lay on a team of professional plasterers to do the job properly for you. Not only will they be able to work much faster than an amateur, which is important, but they will produce a far better job.

There is no fixed rule concerning the size of the labour force. But the more there are the better. During the plastering of a 65 ft yacht hull we had a work force of seven professional plasterers and about 30 other helpers who were used to mix the mortar, deliver it and apply it to the hull. The sole job of the plasterers was to work on the outside of the hull and provide the finishing. This is a time when you find out who your friends are. During the plastering of another 63 ft hull a work force of about the same number was on hand.

You can work out roughly how many hands you will need. There should be at least two people at the mixer preparing the mortar. Then there must be those who will deliver the prepared mortar to the hull. In practice it has been found that a professional plasterer is capable of properly handling about 200 sq. ft of surface area, or perhaps slightly more. He will then need at least two helpers on the inside who will be applying the mortar through the membrane. For a 40 ft hull we believe that at least four plasterers will be necessary with three assistants to each.

Humidity and temperature play an important role in the plastering of a hull for they affect the set of the mortar. For plastering the best possible day would be one in which there is an overcast sky, with the humidity fairly high and the temperature cool. Don't attempt to plaster the hull if it is raining, unless you happen to be working under cover. The downpour will seriously alter the water content of the mortar and the newly plastered hull will be pitted with rain splashes. The rain could also leach out the cement from the implaced mortar, causing it to lose its strength and bonding properties. If the weather happens to be particularly warm, it might be necessary to introduce a retarding admixture to the mortar. Or the water temperature can be cooled to decrease the setting rate of the cement. One idea is to drop blocks of ice into the water to cool it prior to mixing. A number of sprays can be set up round the plastering area to cool the atmosphere and increase the humidity.

Plastering day

On the big day on which you are to plaster the hull, get an early start so that you have the whole day ahead of you. Begin by briefing your labour force. Try and impress on those who are to carry out the mixing of the mortar the need to accurately maintain the fine limits of the mix design, particularly in relation to the water. On no account should an excess of water be introduced to the mix. To the others stress the importance of obtaining maximum penetration of the mortar through the steel membrane. Mortar should only be applied from one side of the membrane and forced through to the other. Mortar which is partially set or relatively old should not be rejuvenated with more water and should not be applied to the hull. You will find that it will not bond properly with the other mortar and causes weakness. It should be discarded totally. If a mix is made too wet, it might be better to discard it as well and start over again. Adding more cement or aggregate in an attempt to change its state will not provide a consistent mix and so should not be attempted.

At the start it is good to have a trial run first to check that the composition of your mortar is just right, and to gauge exactly how much water is going to be needed to achieve the right degree of workability. It will also give your crew of helpers a chance to see exactly how the mortar must be applied. For this a couple of test slabs should previously have been fabricated, conforming to the same design as the hull membrane. These slabs can later be used for strength analyses, if necessary.

Now to the mixing: from a practical point of view it is a good idea to prepare your batches working from full or half bags of cement, as these

34 Feverish activity during plastering day. Note the wooden rule fastened along the deck sheer, enabling the plasterers to trowel up to a straight edge.

give you a consistent weight measurement and a fairly consistent volumetric measurement. Only mix as much mortar as can be handled at one time. Mortar standing by until used will just continue going off, giving you less time to work it into the reinforcement before it is unusable.

There are different ways of adding the ingredients of the mix. But first get the mixer going. It is a good idea to add a portion of the mixing water first; then add the pozzolan or whatever admixture you may be introducing to the mix; follow this up with the sand and cement. Pour in the constituents slowly to enable them to become thoroughly mixed with the water, preventing any dry pockets forming. Once all the sand and cement is in keep turning and add the rest of the water a bit at a time, and check the consistency and stiffness of the mortar all the time. Note that the state of the mix will change all the time from initially appearing very dry to the consistency required. So take care when checking the mix not to be mistaken. Reach in and grab a quick handful without stopping the drum. Squeeze it through your fingers: you'll soon get to know whether it is right or not. Don't over-mix the mortar. The continual turning over of the mortar might accelerate the set slightly and will merely air-entrain the mix.

Plastering

The main object during plastering is to force the freshly mixed mortar into every conceivable space contained within the steel membrane so that when both surfaces are eventually trowelled smooth you will have a dense impermeable shell. Penetration must be absolute, though in practice this is hard to achieve. The surface layers both inside the shell and outside must be trowelled smooth and rubbed down to provide a thoroughly good finish.

There are two basic approaches to plastering a boat hull. One involves plastering it in one single operation, forcing the mortar through the mesh from one side and smoothing it off on the other as it appears. This operation is a continuous one without any breaks occurring. The second technique is variously described as the two-shot method, or the two-stage technique. In this the outside of the hull is plastered first with the mortar being forced only part way through the mesh membrane. After sufficient curing the interior is then plastered and finished off.

Of the two techniques the first can possibly be described as the traditional method. And it is undoubtedly the most popular technique employed by most ferrocement boatbuilders. Until recently, most ferrocement pundits have stressed the importance of completing the plastering operation in one continuous day's work. This instruction stems from

35 When the mortar is forced through the mesh reinforcement it is trowelled thinly against the outside by the plasterers. Here the grid outline of the mesh is clearly visible through the plaster. A thin surface coat of plaster is applied later to finish off the hull and provide a smooth surface.

conventional concrete practice where a continuous pour is always advised, for cold joints are looked on with disfavour. A cold joint is where old cured mortar meets fresh mortar. These are potentially weak points in the structure and should be avoided if at all possible.

The importance of plastering the hull in one go has always been stressed. But opinions have changed somewhat and the feelings now are that the dangers of cold jointing have been overstated, especially where ferrocement structures are concerned. The high steel content corrects or at least counters most of the evils associated with cold joints, but introduces another problem which is that of penetration. This is a much more severe problem than is ever found in normal concrete practice. It takes time and a lot of effort to achieve good penetration so that even with unlimited manpower it is felt that it is often better to break the operation up into separate phases. Hence the development of the two-stage technique, which some builders feel is a much safer and far more practical method for those who have limited labour facilities or where the hull may be very large.

Certain builders now condemn the one-shot or continuous pour method, labelling it as an armchair method put out by theorists. They feel that in theory it is obviously to be recommended but presents many difficulties and dangers in practice. The labour force are generally pressured into finishing the job in one go with the result that they hurry the operation. In the rush quality of workmanship is lost, and insufficient time is allowed to concentrate on achieving maximum penetration. The danger of voids being left within the structure becomes greater.

Richard Hartley wrote to me, relating the experience of an associate of his who built a ferrocement boat employing the one-shot technique, believing it to be the only answer. This fellow engaged a team of professional plasterers who had at one time and another worked on 30 other hulls. Together they plastered the hull and, despite their previous experience, he still discovered 36 large voids in the shell. It took three months working at weekends and at other odd ocasions to locate each of the voids, bore holes into them and fill them with a cement grout. The hull was a 36-footer designed by Herreshoff.

A number of similar stories have been related to me which illustrate the inherent dangers of the one-shot technique, when due consideration has not been given to achieving good penetration. Nevertheless many concrete hulls have been successfully built in this way so the technique should not be condemned out of hand. The general feeling among many experienced builders here in Britain is that where the hull dimensions are not large and you do have a large work force to back you up, with many of them experienced plasterers, the one-shot or continuous pour technique is to be recommended. But where a large hull is to be plastered the two-stage technique is the better proposition. In the first place the work can proceed at a more leisurely pace and the operators can concentrate more fully on the job of obtaining adequate penetration and the elimination of any likely voids. More time can also be spent by the plasterers in getting a good finish to the surface of the hull. A further advantage, particularly from the amateur's point of view, is that labour costs can be reduced, as a large labour force will not be as necessary. Now let us take a look at each technique in turn.

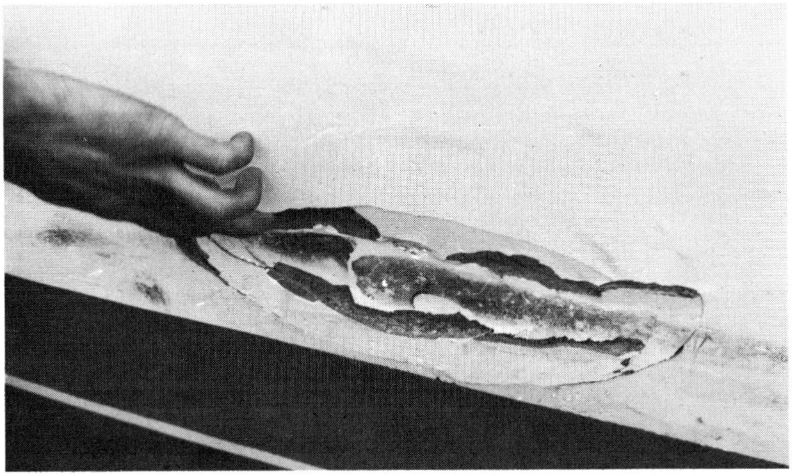

36 Voids will seriously weaken a hull, as is amply illustrated here. Voids were discovered when this ferrocement yacht collided with a tug. Top picture shows the apparent damage at the point of impact. When all the loose plaster was knocked away a large void was revealed in the hull, where mortar hadn't penetrated the mesh during plastering, as can be seen in the bottom picture. The damage was easily repaired with an epoxy/cement filler. Had the void not been there it is highly probable that the hull would not have sustained any damage at all.

Single-stage plastering

This method involves plastering the hull in one continuous pour, the plaster being applied from only one side and smoothed down as it appears on the other There is no fixed rule about which side the plaster should be applied to, but most builders apply the mortar to the inside of the shell and force it through to the outer skin. Always start from the bottom or lowest point of the hull and work upwards. As the hull is standing upright start at the keel.

Begin by applying the mortar to the base of the keel, forcing it through the mesh as best you can. In this area you can use your feet to pound it in. On the vertical sides of the keel and the rest of the hull do not exert your full weight for you might bend the reinforcement out of alignment. This initial application constitutes the structural mortar in which penetration of the reinforcement is the prime consideration. Impress this fact on the operators, but it would be a good idea to closely supervise this early phase. Hired labour, or even friends, if they are not reckoning on sailing with the vessel, will be inclined to get careless, particularly after a couple of hours of concentrated effort. The job of applying the mortar and forcing it through the mesh requires not a little effort.

Operators should work simultaneously on both surfaces of the shell, a labourer applying the mortar from the inside and the plasterer on the outside trowelling the mortar as it appears, skimming it down to the outer layer of mesh. The best way of applying the mortar, unless you are using a mortar pump, is collect a large dollop in gloved hands and cast it with as much force as possible onto the mesh. Then punch and force it through with your gloved fist. Work on the instructions of the plasterer outside who should tell you when adequate penetration is achieved, indicated by the appearance of the mortar at the outer surface of the hull. Avoid using a trowel or any other similarly shaped tool for this job, as it will tend to distribute the effort of the arm over a wider area. The mortar will be relatively stiff owing to the low water content and can be worked into the mesh better by hand than with a trowel. Continually stress the importance of obtaining complete penetration.

The vibrator will aid penetration, so use it. Lay the prod against the reinforcement so that the vibrations will be transmitted through the hull. In areas like the keel where the mortar will be in larger quantities, the end of the vibrator can be stuck into it for better agitation. With the reinforcement in a localised area round the prod vibrating, any air bubbles within the mortar will be dislodged, rising to escape at the surface. The agitated mortar will flow into the spaces which have been left, compacting into a denser medium. In theory, and in practice in concrete work, vibrating too long in one area could cause separation as the aggregates are caught up around the reinforcement. This is unlikely in ferro, as the aggregate particles are smaller. Separation must be avoided, as a loss of strength will result.

In areas where there is a high concentration of reinforcement, extra effort must be made to ensure adequate penetration. It is easy to leave

37 Considerable force is needed to force the mortar through the mesh membrane and achieve complete penetration. Here workers are using their feet to push the cement mortar through the mesh at the turn of the bilge. Care must be taken that the reinforcement is not bent out of alignment in any way. Note the prefabricated athwartships gussets which will be used to support the cabin sole.

38 In the single phase plastering technique, when the hull is plastered in one operation, the mortar should be applied from one side only. Usually it is more convenient from the inside. Here two operators can be seen forcing mortar through the mesh using gloved hands. A dollop of cement mortar is pounded through with the fist until maximum penetration is attained. Plasterers on the outside can then trowel the excess smooth.

voids in such spots as it is difficult to move the mortar properly round the reinforcement. Such areas are around the frames and the prefabricated webs for the later attachment of bulkheads, probably around the sternpost, and alongside any skin fittings, like chainplates or the shafts for the rudder post or propellor.

Go back over sections already worked to check that all is in order. Use a flexible batten to check the fairness of the hull. It is possible the reinforcement may have been pushed out of alignment during the application of the mortar. Now is the time to remedy any deflections by shoring up the mesh. It will be too late to think of doing it once the mortar has had time to set. On occasions you might discover a slight sagging of the mortar, which could mean that there is an excess in those areas. Do not attempt to force it back but rather scrape it off, right back to the wire. This applies inside as well as on the outside of the shell. If the mortar is still soft it can possibly be used to carry the work forward, but don't try to retemper it with more water if it has appeared to have stiffened slightly. Throw it away instead.

In such spots as the bulkhead frame, or interior fabrications like the tank framing or engine beds, which might not be plastered at the same time as the hull, halt your mortar line a couple of inches away. Do not bring the mortar over the fabrication or right up to it. This particularly applies where only starter rods have been left, with the fabrication to be constructed later. There must be good access to both inside and outside faces of a cold joint when these areas are plastered later. The joint will have to be treated prior to the mortar being applied. This is often done by dosing it with a cement grout, or sometimes an epoxy which has been flexibilized by the addition of a polyamide or polysulfide. The fresh mortar is then introduced and worked along the joint for better bonding.

The task of delivering the mortar and forcing it through the mesh is a tiring one. A fatigued person will lack the concentration and necessary effort to continue ensuring that proper penetration is achieved and no voids are left in the shell. So as the plastering progresses more supervision might be necessary. It is a good idea to try and arrange your work force in shifts, so that a couple of hands can be allowed off for a brief rest or refreshment before getting back to the grind, much rejuvenated. But always keep some labour hard at it.

Once the whole hull has been impregnated with mortar leave it for a period until the plaster is just beginning to stiffen, but able to be worked with ease. Run a trowel over the surface and if it maintains its form then the finishing and smoothing off stage can begin. Don't wait too long. Once the mortar reaches a state in which it just begins to crumble, you will know things have been left just a little too long. For the finishing the mortar must still be workable.

This job of laying a final mortar skin over the hull is primarily the job of the plasterer. It must be done with great care and the expertise with which it is done will determine what the hull finish will look like. Only a very thin layer of mortar must cover the mesh, just sufficient to protect it. If possible it should not be more than $\frac{1}{8}$ in. thick. Any extra cover will only add unnecessary weight to the hull, contributing virtually nothing to the strength. An extra $\frac{1}{8}$ in. thickness of mortar on a 60 ft hull could add about 3 per cent to the total displacement. With a hull weight of around 40,000 lbs this could mean a weight increase of around 3,200 lbs. Also, if the external layer of mortar is too thick there will be a tendency for surface fractures occurring as the mortar dries out and shrinks. They will only penetrate as deep as the outer layer of mesh and will not affect the strength of the hull in any way, but they will have to be covered up and filled before the hull is painted so that a smooth surface is presented.

This outer layer of mortar is trowelled on by the plasterer, working it thinly over the surface. A wooden trowel or float is preferable to a steel trowel for this job. If the latter is used it will tend to draw the water and cement to the surface, thus causing segregation and a reduction in strength of the mortar. As the hull was earlier plastered from the bottom, start again from this area, for the mortar will have begun to set earlier here than around the top. As the plasterer completes each section the flexible batten should be used to check the fairness of the hull and spot hollows or bumps. These must either be filled in with more mortar or any excess scraped off. Don't re-use any of the mortar scraped off unless it is still fresh. Dump it. Where slight bumps are discovered in the shell, you could attempt to force the mesh in to remove them, but this must be done with extreme care. If the so-called structural mortar has already gone off slightly, which one hopes it hasn't, disturbing it will create micro-cracks. But this should not occur during this phase unless you have waited too long. The odd wire tie or other loose ends of steel may be discovered sticking out of the mortar at this time, forced out during the application of the mortar. These must be carefully bent back into the membrane, or snipped off. This stage of the plastering is important for the finish of the hull, so frequent eyeball checks would do no harm, so that any unevenness can be spotted and rectified.

With the outer layer of mortar plastered in place a similar procedure can be carried out on the inside of the hull. Of course it need not be as well

39 Plasterers put the finishing touches on the 63 ft ferrocement hull built by amateur Chris Cole, at Brentford, West London.

finished as the exterior, for much of the shell will be covered by lockers, bunks, insulating even, or other paraphernalia. First any surplus mortar must be taken off and the hull scraped down to the mesh. Dump the scrapings. Trowel on a thin layer of mortar just sufficient to cover the mesh. Again ensure that it is not applied too thickly, though this does not matter all that much below the waterline, where it can form part of the static ballast. In areas where ballast will later be placed, like the keel, there is no necessity to put on a surface coating of mortar. Leave it bare. The roughness will help to key in the ballast and improve the bond.

While the inside is finished the surface layer on the outside of the hull has been given time to set slightly. It is possible that with sufficient plasterers both sides can be worked on simultaneously. When the thin

40 Interior view of a ferrocement hull being plastered. Workers are scraping off the excess mortar on the inner surface of the reinforced shell. Mesh knees at the top of the picture will be plastered when the deck structure is fabricated. In this case a narrow ferrocement flange was constructed to which wooden beams and deck were bolted.

plaster coating on the outside of the hull has set to the point where it is crumbly to the touch the final phase can begin. Trowel marks and other imperfections must be smoothed away with a damp sponge worked with a circular motion over the surface. In some cases it is a good idea to coat the hull with a very fine cement grout, applied first with the whitewash brush mentioned earlier. The grout is flicked onto the surface and then trowelled smooth. The sponge can follow up the grout. The hull is now complete and the curing process begins.

Two-stage plastering

The two-stage technique is similar to the method described above except that the mortar is not forced completely through the mesh but stops at or just inside the furthermost layer of mesh. The second stage is hardly more than a smoothing coat of mortar to cover up the steel inside the hull and provide a reasonable finish.

The mortar is plastered onto the outside of the hull first and forced through the reinforcement until it just reaches the inner layers of mesh. The method of application is the same as for the one-shot technique and must be completed in one continuous operation. Again start from the lowest point of the hull and work up to the deck. If a concrete deck has been laid, or the reinforcement for it, then this can be plastered at the same time. The underside of the deck is plastered first, with only a thin layer of mortar penetrating the matrix. Anything much thicker and you will find that the mortar will sag under its own weight. (Where the one-shot method is adopted and a concrete deck is to be incorporated, the deck can be plastered at the same time, in the same way as described above. But most builders adopting this technique leave the deck until a later date as they are usually hard pressed just to finish the hull.)

After the first mortar application in the two-stage method has been completed and the surface smoothed over in the usual manner, the hull must be left to wet cure for at least 10 to 14 days. Before the second phase can begin the inner surface of the hull must be properly cleaned off and all loose material removed. It must then be brushed down with a cement slurry or grout, followed up immediately by the second layer of mortar. Again apply the mortar from the lowest part of the hull and work upwards. Place the mortar with your hand adopting an upward motion. This will prevent any air being trapped within the structure at the cold joint. Adopt this method throughout the application. Instead of the cement grout an epoxy bonding compound could be used to ensure better adhesion between the old and new mortar. When forcing the mortar into the mesh be careful not to push too hard. The old mortar although set will not yet have had time to develop full strength and any movement caused through pressure could create cracks. These will only be minute and will heal at least temporarily by the deposition of lime salts during the wet cure. I say temporarily, for if this lime deposit is exposed to any water it could leach out. One of the theoretical advantages of pozzolan is that it will react with the free lime while in the presence of moisture to form an insoluble calcium silicate with good cementing qualities, thus effecting a permanent repair.

With the hull freshly plastered make every effort to ensure that no movement takes place as this will do nothing but harm the hard work that has been put in. Note that where the props are supporting the hull, gaps will have been left by the plasterer. These will have to be filled in after the hull has cured sufficiently for the props to be shifted to a different position. On no account attempt this job while the hull is still green. A stiff epoxy filler (resin built up with neat dry cement or industrial sand) will provide a good hard patch. The hull will need further finishing to bring about a really smooth surface, but this is best left until the mortar has properly cured. Only if prominent imperfections are noticed soon after plastering has been completed should any attempt be made to correct them. Ridges and lumps can be ground down after the hull has been left for 24 to 36 hours, but this must be done with extreme care. Use some coarse carborundum paper, or even water paper to achieve the necessary effect. The mortar will wear away easily while still in a green state, so work slowly and try not to expose any of the reinforcement. This preliminary grinding should only be attempted if it is absolutely necessary. Much better to leave it until the hull is settled and in a reasonably hard state. Now the essential thing is to leave it well alone and allow the curing process to work.

ferrocement 11
CURING

Proper curing is essential if the desired properties of the mortar are to be developed fully. The water introduced to the mixture of aggregate and cement reacts with the latter to form an insoluble compound, calcium silicate, which binds together the particles of aggregate into a hard matrix. Only small quantities of water are necessary to initiate this reaction, the process being described as hydration. Although initially rapid, it is a slow process which will continue only as long as there is moisture present within the mass; full strength and hardness may not be realised for years. Curing control ensures that the process of hydration continues, and at a constant rate, which is important. It maintains the presence of water, vital during the early life of the mortar, and regulates the temperature level, which can also affect the rate of strength development.

Ferrocement structures, because of their thin shell-like form are prone to evaporation, which must be prevented or it will cause premature drying out of the mortar. Early drying out could remove the moisture from the mortar which is essential for hydration, thus terminating the strength development. Great shrinkage will occur during drying with severe stresses developing as a result of the volumetric change. If shrinkage occurs too rapidly, too soon, before the mortar has obtained adequate strength, the material will fracture and cracks will result. So proper curing during the early stages after the hull has been plastered is especially important. High humidity must be maintained to ensure that evaporation does not occur.

Hydration of cement is described as an exothermic reaction. Changes of temperature will affect the rate of hydration: increases in temperature will accelerate it; a reduction in temperature levels will decrease the rate of hydration and hence the rate of strength development. In normal curing conditions ideal temperature levels would be in the order of 50°–90° F. However, the mortar can be subjected to much higher temperatures, perhaps around 180° F, to increase its rate of strength development, but only providing that no premature drying occurs. This is often achieved by low-pressure steam curing. Low temperatures will slow the curing of the mortar and if it drops as low as freezing, as it could well do in winter conditions, hydration may even be halted. Basically, good curing involves keeping the cement mortar continually moist for a minimum of seven days and preferably for at least 28, while maintaining a constant temperature level around 70° F. During this time the hull should be protected from hot dry winds which would encourage rapid evaporation.

Accompanying this section is a table which illustrates the compressive strength development of cement mortar at different times for different conditions of curing. The figures were supplied by the New Zealand Portland Cement Association and relate to concrete which gains compressive strength of 4,500 lbs per sq. in. after 28 days moist curing, this level then further increasing to 5,800 lbs per sq. in. after 90 days. From it we can see that seven days' moist curing would be the practical minimum if ultimate strength gain is to be comparable with mortar curing after 14 to 28 days. A lot of strength will be sacrificed if moist curing is terminated below the seven day period, while an extension of the wet cure can only improve the quality of the mortar. Note that the long-term strengths of the samples cured continuously in moist conditions differ but little from those which were allowed to dry after the 28 day 'incubation' period. At this juncture it should be pointed out that the other properties

Table indicating compressive strengths of concrete in lbs/sq. in. for periods of moist curing (prepared by New Zealand Portland Cement Association)

Moist curing—no. days continuously moist cured	*Age in days* 3	7	14	28	90	180
Continuously moist cured	1700	2500	3700	4500	5400	5800
28 days	1700	2500	3700	4500	5700	5700
14 days	1700	2500	3700	5100	5100	5100
7 days	1700	2500	3800	4600	4600	4600
3 days	1700	2600	3400	3800	3800	3800
No moist curing	1500	2000	2400	2500	2500	2500

of the mortar, such as impermeability, durability, etc, are related to the compressive strength and will show similar rates of development.

Ferrocement boatbuilders adopt a variety of techniques for curing their newly plastered hulls. Generally they conform to two patterns, using either what can be called natural curing procedures or low-pressure steam. Natural curing is the technique followed by most builders, particularly amateurs who do not have the facilities for steam curing.

Natural curing must encompass the complete 28 day cycle (it can be carried on longer) during which time no further work should be carried out on the hull. This method necessitates keeping the mortar, inside and out, continually moist to prevent drying out and shrinkage. Wetting must be done at regular intervals and over a period lasting not less than 7 to 10 days. It should not begin, however, until about 24 hours have elapsed since plastering, though this limit is somewhat arbitrary. If sufficient time has not been allowed for the preliminary set of the cement, wetting down of the hull can cause separation of the mortar, leaching out of the cement, and subsequent reduction in strength.

The hull should be covered with some sort of moisture barrier so that a constant high-humidity atmosphere can be maintained. It will also act as protection from wind and sun, which accelerate evaporation. The usual method is to envelop the hull in plastic or polythene sheeting and then cover the mortar inside this covering with a continual fine water spray. Before watering prepare small drainage holes at the bottom of the keel to allow the curing water to escape. These holes can be placed prior to plastering if need be.

The total surface area of the hull must be kept moist. The spray must not be hard or concentrated in any one spot for this will only damage the hull. One way of applying it is to drape perforated hosing all over the hull inside the plastic coverall. Let this run continuously over the inside of the hull and over the edges. Or the hull can be dosed about every two hours, the intervals to be determined by the humidity level and the temperature. If it becomes really warm and there is a danger of evaporation occurring, then spraying must be carried out more frequently. But it must be regularly done. Alternate wetting and drying of the mortar will introduce a certain amount of expansion and contraction which could well result in cracking and fracturing of the surface.

Hessian sacking and cotton waste are often used in addition to or in place of the plastic sheeting. However, for it to be effective it must be wrapped fairly tightly round the hull and must not be allowed to flap around in the breeze. Cotton waste packed in thick wads can be useful for the edges of the hull. Properly moistened it will distribute the water over the surface and prevent any localised evaporation. It will not be sufficient just to occasionally spray the hessian cover with water, but it must be thoroughly soaked so that the hull will be completely moistened.

If the weather is cold during this critical curing phase, the low temperatures will reduce the rate of hydration and strength development will be retarded. Bear this in mind for it may be necessary to extend the period of cure to ensure that strength development reaches the necessary level before any work is carried out on the hull. Hydration can be assisted in such conditions by warming the water being sprayed on the hull, though this might present a few practical difficulties. One builder suggests heating buckets of water over a couple of hotplates placed under the moisture cover. The evaporation from this would raise the humidity temperature level. However, the inherent dangers of this setup appear to outweigh any advantage which may be derived from its use. If the hull is in a shed blower heaters can probably be used to raise the air temperature slightly without disturbing the humidity level. Gas heaters should not be used at this stage for the products of combustion could possibly affect the curing process.

Membrane-curing compounds, which are sprayed on the outer surface of the hull, are being used more often here in Britain as a sealant to prevent the premature evaporation of moisture from the mortar. They are used in place of the plastic sheeting, although it is still good practice to cover the top of the hull with some sort of barrier, like canvas or plastic, and the inside of the hull with hessian or cloth. These compounds have filled their role efficiently and effectively, but certain types may present problems later when the time comes to paint the hull. Most of these compounds are designed to break down under the influence of sunlight about a month after application, when the mortar has cured sufficiently. Consequently they should not be used where the hull may have been built in a shed or under any other type of shelter which will act as a shield against the sun's rays. It is recommended, too, that they only be adopted where there will be a delay of a few months between plastering and painting of the hull. Although this should not be an obstacle as it is not advisable to paint the hull until the mortar is completely dry—this will take a couple of months at least. A curing compound which has been successfully used on a couple of concrete boats here in Britain, named Rite Cure, has presented no problems when it came to painting. Six weeks after being applied to the hull it could be seen to be breaking down and flaking off. A hard brush helped the process along.

If membrane curing compounds are to be used they should be applied as soon as possible after the hull has been plastered, and to the outside only. The application of such a compound does not mean that there is no need to continue keeping the hull moist with a water spray. The interior of the hull must be moistened continuously, especially during the first seven to 10 days after plastering. The sealant can be applied with a sheepskin roller, or by spraying through one of those hand-pump insect sprays, after it has been thoroughly cleaned of its original liquid. Plant sprays, with an elongated cylinder carried in a back pack, have sometimes been used here. A brush is not recommended as you are likely to damage the surface mortar, or at least leave brush marks in it.

As has been previously mentioned, an effective method of accelerating the strength development of the cement mortar, without the danger of premature drying out, is by curing with low-pressure steam. Wet steam is used to create temperatures of up to 180° F while maintaining a constant moisture level. This has the effect of increasing the rate of hydration to the point where strengths normally developed after 28 days by natural curing will be attained in about 24 hours. This offers obvious advantages to the commercial builder in that it shortens the lay-off interval necessary before further work can be carried out on the hull. Another advantage over the long-term method of curing is that it eliminates the extended periods of water spraying which can seldom be carried out regularly without some interruption. However, it should be noted that steam curing gives mortars with slightly lower ultimate strengths. Few amateur builders have to my knowledge yet adopted this faster technique, apparently for no other reason than a lack of facilities.

The technique involves shrouding the hull in a close-fitting cover, which can be made out of polythene sheeting, or canvas, or anything else which can be insulated if necessary and will act to retain the steam close to the surface of the plaster. Some sort of framing will be needed to support the coverall and hold it away from the hull. The steam will then be piped in under it. A steam generator is used to produce the low pressure steam, and is similar to those machines which are sometimes used to clean the outside of buildings. The piping system must be so laid out as to allow the steam to reach all parts of the hull uniformly. A series of perforated pipes spread out around the hull and connected to the generator is a simple answer. It is important that the steam is applied constantly over all surfaces with no gaps anywhere which could give rise to temperature differences, which can cause expansion and contraction of the structure. This would be particularly detrimental while the hull is still in such an early state of set.

A serious effect would be a slight degree of hull fracturing. More lasting would be a 'non-uniformity' of strength throughout the structure. In practical terms it is extremely difficult if not virtually impossible to obtain a totally gradient-free atmosphere round a large hull, but every attempt should be made to do so.

Steam curing should not commence until some time (about 24 hours) after the initial set has taken place as the wet steam could cause leaching out of the cement. The process must be carried out gradually for sudden temperature rises are to be avoided, and the rate of temperature increase should be maintained uniformly over the whole hull. The recommended rate should be in the region of 30°–50° F per hour, which means that it would take about 4 to 5 hours to achieve the maximum 180° F level. From then on steaming must continue for 6 to 9 hours. However, this curing cycle will depend to some extent on the characteristics of the constituents in the mortar, the mix design and the temperature conditions prior to steaming, so these curing times could vary slightly. After steaming the temperature level should not be allowed to drop suddenly, but only gradually over a period of 4 to 5 hours. Steaming periods or cycles may safely be extended, but they should not be shortened as strength development would be terminated before the ultimate level had been reached.

ferrocement 12
FINISHING AND PAINTING

The finishing layer on a ferrocement hull should provide a cover not more than between $\frac{1}{16}$ in. to $\frac{1}{8}$ in. over the steel mesh. The skill and patience of the plasterers in applying this thin outer covering of mortar will determine how much finishing work is necessary before the hull can be painted. You may take it from me that it is far better in the long run to provide a good finish initially than to attempt to achieve one later, after the hull has cured. Not only will you save hours of labour, but you are likely to have a far better looking hull in the end.

Once the mortar has cured the main job of rubbing down and smoothing off can be carried out. High spots left earlier can now be ground off and all the lows filled in. No matter how carefully the plasterers finished off there are bound to be a few imperfections here and there. Carborundum stones can be used to grind down the hull. A machine we have used to good purpose for this job has been one of those industrial rotary disc sanders with a dense felt pad to take a carborundum disc. With very little practice the sander will work well, floating over the low spots and taking off all the bumps or ripples left in the surface. Be careful to keep the disc flat against the hull or you may end up digging a rut in the mortar. Work slowly and be careful not to grind the mortar completely off, exposing the mesh. We have found it good practice to take the sander over the hull at least once before attempting to fill in any hollows there may be.

Low spots, pores, etc. can be filled in with a stopper. Before this is done, the hull must be completely cleaned off and all the dust particles created by the sander removed. It is possible that a fine powder or latience may have been formed during the curing of the hull. The laitence is composed of a weak, chalky coloured deposit of cement and aggregate which may have been flushed to the surface of the concrete. It is a sure sign that there was an excess of water in the mix, and that curing may have taken place too rapidly. This dust coating can be scrubbed off with a brush and fresh clean water, or the hull can be washed down with a weak acid. Muriatic (hydrochloric) acid has been used satisfactorily. The acids will also key the surface of the hull, making for better adherence of the paint later on.

The hull must be completely dry when the filler and other protective coatings are applied. It must therefore have cured entirely. When it is dry it will look dry, becoming a lighter shade of grey. Any dampness will just ruin the bonding and it will not take long for the paint or other protective coatings to flake and peel off. There are a number of different materials which can be used as fillers. One of the most satisfactory we have found is an epoxy resin based stopper known as Epoxy Solventless Stopper, manufactured by the Little Ship Division of British Paints Ltd. It is a two-pack resin which comes in 1 pound units. The filler is applied with a spatula, trowel or other similarly shaped tool. Once stopper has been applied and set it will again be necessary to turn the sander or carborundum stones onto the hull, to blend in the filler with the rest of the hull. A slightly finer grain paper is recommended at this stage. The stopper is not as hard as the cement mortar, so extra care must be taken when sanding not to destroy the good work that was done before. When the really rough spots have been removed it is best to employ very fine sandpapers. A wet and dry paper will provide a fine finish. This operation can take quite some time if you have many imperfections to cover up. It is a tedious job but worth it in the end if you are looking for a good finish. The hull must again be washed down with fresh water after this grinding and sanding is complete.

Allow the hull to dry off again before attempting to apply anything else to the hull. It will regain the pale colour of dry concrete. It is not absolutely necessary to paint the hull with any protective coating. The cement mortar should be virtually impermeable, and durable. But it is an added safeguard, and anyway a painted hull is more aesthetically pleasing than a large expanse of grey.

Epoxy based paints have proved to be the most successful for ferro-cement, bonding well to the cement mortar and giving greater protection for a far longer period. The procedure which should be adopted is to coat the hull with a compatible primer or sealant. Material which is often used for this is an epoxy-tar compound, which is a black two-pack resin. An ordinary clear epoxy resin will also do just as well, while another primer which is sometimes suggested is chlorinated rubber paint. Adhere strictly to mixing instructions supplied by the manufacturer, for proportions of base and hardener are quite critical. Also look to the environmental conditions. Avoid extremes of temperature and attempt to apply the resin, and the later coatings of paint, before evening moisture descends. Avoid very humid conditions. And, lastly, make certain that the surface of the hull is clear of dust, grease and other dirt. Clean surfaces are essential if good adherence is to be achieved.

The primer is a thick, heavy material and not easy to apply evenly. We have adopted various methods for applying it. A screed can be used to spread it over a wide area, followed up by a brush and roller. Once the resin base has set hard, which could take 24 hours, it will probably be necessary to once again sand down the hull, to remove any bubbles and lumps which may have been left. Use a very fine grain paper.

Now to apply the top coat of paint. Follow manufacturers' instructions in determining the number of coats. Usually two are quite sufficient. Epoxy paints are found to be superior for ferrocement craft. A good bond will be achieved, particularly as the primer is an epoxy as well. Polyurethane paints have given good service, painted onto an epoxy base. Vinyls and acrylics have also been successful. Enamels should be avoided as they will show up any tiny imperfection in the hull. A matt finish is much better. It is not necessary to paint the bottom of the hull. It will not be seen and will only be a waste. Antifouling, which is normally applied shortly before the hull goes into the water, can be applied directly onto the dried and cured ferrocement, but it is better practice to have an epoxy base to adhere to. The antifouling will remain on for a much longer period, and repainting will require less work. Paints can be applied with a roller or fine brush. Take care to clean off the tools, brushes and rollers after painting or they will be completely ruined.

ferrocement 13 CONSTRUCTION DETAILS

Keels
Bulkheads
Tanks, engine mounts, shaft installation
Decks
Fittings
Rudder
Repairs

The following elements must be considered at the design stage as they involve the framing and longitudinal reinforcement and are stressed.

Keels

The keels of ferrocement yachts can be constructed in various ways. The most common and perhaps most practical method is to integrate the keel with the hull shell, developing a trough section into which the ballast can then be fixed. This trough keel is fabricated with the hull station frames extended down to include the keel.

Another variation is to precast the keel with the heavy aggregate or ballast imbedded in the mortar. Steel lugs are left protruding from the top of the keel for the attachment of the station frames and the hull reinforcement is then married to the keel, a cold joint being formed at the keel/hull interface. This technique offers the advantage of a firm and rigid base on which to erect the hull fabrication, reducing somewhat the dangers of deflection. This idea has been adapted by some builders using the trough keel section in which the keel was laid up and plastered before continuing with the rest of the construction. This provided an anchor for the frames, restricting any movement while the stringers and mesh were being fitted.

A third idea is to use a precast lead keel bolted onto the ferrocement hull after the fashion adopted for wooden hulled vessels. In this case the hull round the bilge will have to be suitably reinforced. Plate steel can be used as a backing on which to bolt the keel. It could either be fastened inside the hull, welded to stem and sternpost along the backbone, or be integrated with the structure during early fabrication and welded to each pair of station frames. As yet I know of no builders who have adopted this type of keel assembly.

There are many different heavy aggregates which can be used for ballasting and selection may often be governed purely by availability. Lead, because of its high density, would obviously be ideal as only a small volume would be necessary to meet the weight requirements. But it is expensive and anyway care must be taken when using it in a ferrocement hull. Lead and steel are not good bedmates, particularly when in the presence of moisture, so it is not advisable to bed the raw material in mortar when fixing it in the keel. If it is to be used some sort of insulation would be recommended. One idea would be small diameter precast concrete pipes which could be filled with molten lead, allowed to cool and then packed into the keel. Mortar could then be poured round them to hold the ballast in. Small asbestos pipes would also serve this purpose.

Other aggregates which are not as expensive as lead and which are more often used in ferrocement craft include boiler punchings, steel shot, heavy chain links, steel ball bearings, scrap metal and lumps of pig iron. Concrete is also suitable for ballasting and heavy aggregates may be substituted for the usual sand to increase the density of the mortar. Such aggregates include barite, ferrophosphorus, bauxite or serpentine. There are other aggregates which may also be used but always ensure first that they will not react adversely with the cement or contain any deleterious materials which could impair the set of the mortar. Naturally the best aggregate to use would be that which will provide the greatest amount of weight for the least volume.

Do not place all your ballast before the hull goes into the water, unless you are absolutely sure that your weight calculations, or those of your architect are correct. Put in about two thirds and hold back a third which can be used to properly trim the boat fore and aft and to the designed waterline. If I were you I would not trust your calculations entirely. Take the case of an East Coast builder who believed in the instructions supplied by the designer and concreted in all the ballast prior to launching his newly built yacht. The ballast consisted of steel ball bearings imbedded in mortar. Came the day when the yacht was dropped into the drink for the first time and she was seen to be well down by the stern, her bow poking up into the sky. Some weeks had to be spent chipping out the ballast with the aid of a pneumatic jack hammer and replacing it elsewhere within the hull before the vessel floated as she was meant to.

Placing the ballast and concreting it into the hull is a simple though

41a,b In the construction of a 40 ft TSDY built recently in West Germany T-bars were used to fabricate the station frames instead of the usual black water pipe or mild steel rod. Another variation: the keel was first precast in concrete and used as a base on which to construct the rest of the hull. Steel lugs were cast into the keel and later welded to the station frames of the hull. The yacht was designed by G. L. Watson and built by Ernst Uwe Winteler. 41a shows the T-bar frames of the yacht during early lay-up of the reinforcement. Rod stringers are welded to the frames rather than tied. In 41b the hull has been plastered; the joint between precast keel and reduced trough section of the hull is clearly visible, accentuated by the black line.

back-breaking operation. Take the occasion when we ballasted the 65 ft ketch *Petralis*. Because of the limitations of the crane used to drop the yacht into the water the hull was completely unballasted when launched, weighing at that time only 20 tons. Thirteen tons of ballast were required to bring the yacht down to her marks, the ballast consisting of approximately 3 tons of almost neat cement and 10 tons of pig iron.

During the early fabrication of the hull high tensile steel rods had been placed longitudinally down the keel, linked to the stem and sternpost. After plastering these were left exposed with the idea of keying in the ballast. Now, short lengths of mild steel rod were looped round these longitudinals and led a short way up either side of the keel to further aid the keying of the mortar and restrict any possible movement or lifting of the ballast during a heavy pounding at sea. The pig iron, which was cast in more-or-less uniform shapes roughly 3 in. × 2 in. × 12 in. was packed tightly into the trough keel between the floor gussets. A virtual cement grout was poured in to glue the ballast. It was well tamped to eliminate any voids and completely covered the pig iron, with a slight aft slope to the sump under the engine to allow drainage of any water collecting in the bilges. The mortar was continually soaked to prevent drying out and to assist the curing process. Although almost neat cement was used to hold the ballast and it was fairly liquid when poured, there were few signs after two weeks' curing of any cracks forming.

Bulkheads

Bulkheads can be constructed and attached to the hull in a number of ways. They have been constructed in ferrocement, fabricated and plastered in the same way as the hull or deck; or they have been built in wood or steel plate. Ferrocement bulkheads will greatly improve the stiffness and structural strength of a hull, but because of the weight factor, they are not often included in small vessels and/or are only adopted where structural bulkheads are needed, plywood being used in other areas.

Ferro bulkheads can be built as a precast unit before the hull is built, then erected to fulfill the role of a pair of conventional station frames to provide the mould from which to hang the reinforcement; or they can be fabricated at the same time as the shell, or even after the hull has been plastered. In the case of precast bulkheads mesh and starter rods should be left exposed so that they can be linked to the hull reinforcement. The rods must be bent at right angles so that they can be tied alongside the hull stringers. Rod and mesh dimensions must be the same for both bulkhead and hull. It is important that the joint between bulkhead and hull is strong and so designed that any lateral forces on the former will not crack the hull. A precast bulkhead will offer a rigid structure on which to mould the hull, thus reducing to some extent the danger of deflection, but it may accentuate the problems of fairing the hull.

Wooden bulkheads are more usually built into ferrocement yachts, mainly for weight reasons but also because they are aesthetically more pleasing. If a sandwich construction is used, plywood bulkheads will provide adequate strength for structural purposes and impart stiffness and rigidity to the hull. The most satisfactory way of attaching them is by means of ferrocement webs or flanges which may extend roughly 4 in.–6 in. inboard from the hull and run continuously from floor gussets up to the deckhead. The webs are integrated with the hull reinforcement providing strong attachment for the bulkheads with little likelihood of pulling away during the flexing and twisting of the hull. They may be constructed before the hull is plastered or afterwards. Fabrication is carried out by fastening short reinforcing rod sections to the hull reinforcement. These are tied alongside the longitudinals, bent at right angles to protrude inboard from the hull. One or two transverse rods are then welded or tied to them. Steel mesh covers this reinforcement and is wedded into the mesh covering the inside of the hull. Where prefabricated station frames have been used instead of water pipe, this protruding reinforcement is, of course, already prepared. But even then, the accommodation arrangement might necessitate further webs being built in. The webs can be plastered at the same time as the hull, or later. Particular attention must be paid to cold joints, though. A cement grout or epoxy resin brushed over the hard concrete will improve bonding. Plywood bulkheads can be bolted to these flanges once they have cured; arranged as a sandwich construction with ply on either side of the web with fillets of wood separating the sheets; or a single sheet of ply can be bolted to one side of the web. Holes are easily drilled through the flange with a masonry bit. Bolts should be washered or chocked when they are fastened in place to prevent them from biting into the plaster.

Another method of attaching bulkheads to the hull is by welding steel brackets or lugs to the reinforcement prior to plastering. These can be drilled, suitably protected with a coating of epoxy resin to prevent corrosion, and the wooden bulkheads bolted to them. This is not quite as satisfactory as the ferrocement webs. A fourth form of bulkhead attachment is glueing with fibreglass strips. This is a rather flimsy arrangement and should only be adopted where partitions are needed. Bulkheads attached in this way will not provide much strength structurally, for it will not require much force to rip them from the hull.

Floor supports or gussets can be fabricated in the same way as the webs and are usually integrated with them at each station frame. They are best constructed before the hull is plastered and consist of horizontal and vertical reinforcing rods laid up in a grid and joined to the hull reinforcement in the same way as the rod starters for the webs. They are covered with mesh, usually the same number of layers as were used for the hull, and then plastered. Holes can later be drilled to take electrical wiring or for bilge drainage. The top of these supports should be at floor level. Timber cross members can be bolted or fibreglassed on for the attachment of the floor. As a variation to these gussets, some builders have merely bolted angle iron sections to the bulkhead webs to act as floor supports, braced in the middle with an upright.

Tanks, engine mounts, drive shaft installation

Fuel and water tanks can both be constructed in ferrocement, and fixings must be placed for them before the hull is plastered. The tanks are easy to fabricate as their construction is similar to the floor trusses. Where construction is carried out after hull plastering, rod starters must be left protruding to take the tank construction. Provision should be made at the top for an inspection plate, which can be of steel and bolted to the ferrocement. Studs should be provided for this. Baffles will almost certainly be necessary to prevent the liquid in the tanks from swilling about. A simple way of constructing these is to fibreglass strips of ply inside the tank, from one side to the other. They can then be easily removed if the tank has to be inspected, cleaned or repaired. Some builders are against incorporating ferrocement tanks. Their reasons: they cannot be removed and may therefore be difficult to repair and clean. If the hull is damaged in any way repair would present problems. Diesel fuel has a deleterious effect on ferrocement and will seep through it, so bare ferrocement fuel tanks are not considered satisfactory. However, they can be suitably treated to prevent this happening. The interior can be coated with epoxy resin or other protective compounds. Against these disadvantages ferro tanks are inexpensive to construct and can be integrated with the hull.

Engine beds or mounts, if constructed in ferrocement, must be so designed to allow for the proper distribution of stresses through the hull. If necessary the hull should be reinforced in this area, particularly if a large engine is to be installed. Reinforcing rod can be formed to provide the essential framing for the engine beds. The rods should connect to the keel and the other solid reinforcement of the hull so that stresses can be distributed. These rods can be covered with layers of mesh and plastered. Studs can be plastered in, welded to the reinforcement so that the engine mounts can be bolted on. Sections of galvanised RSJ or channel steel can be used for the mounts.

A complicated setup need not be constructed, but it must be strong and rigid. Take the system used on the New Zealand-built yacht *Ahwahnee*. In her case a steel grid conforming to the flare of the hull alongside the keel was welded to the base of the engine bed plates. These were bolted to the engine mounts. The engine, meanwhile, was suspended above the engine well hanging from two chain hoists. Engine and grid were lowered into the hull with the latter pointing down into the bottom of the keel. The flange on the engine drive shaft was aligned with the flange on the propellor shaft. The shaft and stern tube had been installed when the hull was plastered. Once the alignment was spot-on the engine was wedged tight so that it would not move and concrete was poured into the keel around the grid. Apparently this engine mount has worked well. The engine transmits power direct to the prop shaft without any vibration being felt and no shifting of the engine has occurred.

A similar arrangement was adopted for installing an engine on the amateur-built yacht *Petralis*. The engine installed was a big four-cylinder Gardner diesel, originally used in a pick-up truck but converted for marine use. This was done by coupling on a heat exchanger and a marine gearbox. A simple engine mount was built out of channel steel. The forward mount took the form of a horizontal section of steel which was bolted to the engine bearers. This was welded to two channel steel braces, the three forming an inverted triangle. The rear mount, also in channel steel, was made up into a T-section, the vertical directed straight down. These two mounts were bolted to the engine and a fore-and-aft brace was fastened in place. The engine was then lowered into the hull, the supports extending down into the keel. The engine was aligned on the centreline—the stern tube and prop shaft had not yet been installed—and the engine mounts concreted in place.

A simple and effective approach was adopted to install the prop shaft and stern tube. During the initial hull fabrication and plastering an aperture was left in the sternpost just large enough to take the stern tube. This was done by using a section of steel water pipe as a former. Its diameter was somewhat larger than the stern tube. After the hull was plastered this former could be twisted out of the hull leaving a nicely formed aperture.

After the engine had been installed the prop shaft was inserted into the hull and coupled to the flange on the drive-wheel of the gearbox. The stern tube was sleeved on and positioned with the outer bearing sitting flush with the trailing edge of the sternpost. The prop shaft was carefully aligned to eliminate any wobble. This was done by hand-turning the engine and checking with a feeler gauge. The stern tube was wedged tight and cemented in. Instead of a conventional cement grout a filler composed of epoxy resin mixed with dry neat cement was used. Cement was added to the resin until the mix had the consistency of putty, and this was forced into the aperture, surrounding the stern tube to a depth of about 8 in.

42 Adequate mesh from the hull overlaps the deck to be married in with the reinforcement of the deck. Construction of the deck, in this instance, took place after the hull was plastered and had cured.

A two-way thrust bearing was also necessary to prevent the shaft from dislocating. A simple arrangement was devised to install it. A section of channel steel was set up vertically in the keel with braces fore and aft to hold it steady. A hole large enough to take the prop shaft was drilled into the top end of the steel and it was concreted into the keel. The thrust bearings were fastened to the prop shaft on either side of the preventer. The complete engine installation system, from engine mountings to stern tube, has operated satisfactorily with no noticeable vibrations through misalignment or any leakage occurring round the stern tube.

Decks

Decks and cabins can be constructed in ferrocement or in wood. They have on occasions also been moulded in GRP. In most cases where ferrocement is employed it is limited to the deck structure with the cabin built in wood. The main reason for this, which is particularly important in yachts and other smaller vessels, is an attempt to reduce weight. For yachts under about 35 ft in length it is perhaps better to construct the upper works in wood, for the weight of a ferrocement construction will not do much for the stability or sail-carrying capacity of the vessel.

The construction of a ferrocement deck is carried out in much the same way as the hull and may be completed at the same time. But most builders prefer to leave this job until after the hull has been plastered, to allow easy access into the hull while plastering. The unfinished deck reinforcement will not be strong enough to support much weight and so should not be stood on at all. Deck beams can be either of steel water pipe or prefabricated mild steel rod frames and are welded athwartships to each pair of station frames. They must be properly lofted to reflect the camber of the deck. Radius of the camber is usually supplied with the table of offsets.

Longitudinals of high tensile steel rod are laid down the length of the deck at 2 in.—3 in. centres, following the curve of the hull. Their ends can be bent at the stern and bow, overlapping the rod which reflects the deck sheer. Alternatively, and this could occur should the hull be already plastered, the rods can be welded to stem and stern. If need be provision can now be made for toerail or bulwark. Each longitudinal must be wired to the deck beams. Care must be taken not to kink the rods in any way as it will be difficult to straighten them; any imperfections will show up. It might be necessary at this stage to give some support to the deck structure, to reduce the chances of the deck beams being bent out of alignment. Props can be placed beneath each beam, but a more convenient arrangement would be in the form of wire supports, tied to the ends of the beams and up to the rafters or cross bearers of the scaffold.

Six to eight layers of mesh can be used to cover this main reinforcement, the number of layers obviously depending on the gauge and size: follow the design used on the hull. The mesh must be wedded to the edges of mesh left exposed from the hull. Attempt to stagger the joins to avoid an excess build-up of steel in one area. Tie the mesh in tightly with wire ties in the same manner as on the hull. Fair in the deck using a long batten, eliminating any bulges and wrinkles. As no weight should be allowed on the construction at this stage it will be necessary to construct some sort of walkway or scaffold over it. A scaffold inside the hull would also make

43 Teak laminate deck beams on the 65 ft yacht *Petralis*. The beams were spaced at 18 in. centres and fastened to the hull by means of a concrete ring beam. High tensile rods passed through the beams anchoring them to the ring beam. Note the chain plates, made out of galvanised ring bolts and imbedded in the concrete.

work a bit easier. If the cabin is to be built in ferrocement too this can be framed in at the same time as the deck. When preparing the deck fabrication allowance must be made for the cockpit, cabin, hatches, and other deck openings.

When plastering the deck the underside is always done first. This is allowed to cure for 10 to 14 days before the upper part of the deck can be plastered. The reason for this is that the full weight of the mortar will cause it to sag and drop through the reinforcement before it has time to set. A thin layer trowelled on first is unlikely to do so. Some builders consider it worth erecting a shuttering on the underside of the deck and completing the plastering in one go. Where two-stage plastering is adopted the cold joint must be treated with a wet-to-dry bonding agent to improve the adhesion. A cement grout is the traditional technique, but an epoxy resin can also be used. After curing the deck can be finished off and painted in the same manner as the hull. A little industrial sand can be added to the paint if necessary to provide a non-slip surface.

Attaching a wooden deck to a ferrocement hull is not difficult and a true handyman should find few problems facing him. A number of techniques that have been used, of which three are described here. The first method follows conventional wood construction practice. A wooden beam shelf is bolted to the concrete hull and into it the wooden deck beams or frames are let. To bolt on the beam shelf bolts can be welded to the hull reinforcement before plastering takes place. A simpler idea is to drill bolt holes into the concrete using a carborundum masonry bit, which will not take much effort. If a rubbing stake or wooden fender is to be attached to the sides of the hull, the bolts holding the beam shelf can be used to hold it as well. An epoxy can be used as well, to glue the timber to the hull and to seal the bolt holes. The plywood deck can now be glued and nailed to the beams, the edges sitting flush with the top of the ferrocement hull. Any gaps which are left can be stopped up with an epoxy filler.

Chris Cole, an amateur builder who, at the time of writing, is fitting out a 63 ft ferrocement cutter he built near Brentford, West London, adopted a different method of deck attachment. His technique was to construct a ferrocement flange or web, looking something like a narrow side deck, right round the hull at deck level. This was constructed after the hull had been plastered and was fabricated in almost the same way as a ferrocement deck would be. The flange on this particular vessel extended inboard for perhaps 9 in. to 1 ft. After it had been plastered and cured laminated wood beams were bolted to the underside of the web at about 18 in. centres. The beams were prepared on a wooden former to a standard camber. Once these beams were properly lined up and faired in 1 in. plywood sheets were glued and nailed on using Gripfast nails. Where the sheets extended onto the ferrocement web they were glued with an epoxy resin and bolted through. A narrow strip of bare concrete was left round the edge of the deck where a short upstanding toerail was fastened. The deck was sheathed in glassfibre, more for weather protection than for strength.

Another technique was adopted on the yacht *Petralis*. This vessel, which has a beam of 15 ft, is almost completely flush-decked except for a tiny cabin over the saloon. Originally a ferrocement deck was planned but at the last moment this was waived in favour of laminated beams and plywood

44 Stemhead fitting: this was welded to the stem pipe for greater strength.

deck. When this decision was taken a ferrocement deck was already prepared for with mesh and rod starters protruding inboard round the gunwale to knit in with the reinforcement of the deck. Instead of doing away with this steel, it was cut off short and bent inwards to partly reinforce a concrete ring beam which was constructed round the hull. This beam really took the place of the timber beam shelf described opposite.

The concrete ring beam was built by first erecting a wooden shuttering to provide the form. This gave a cavity about 3 in. wide by about 4 in.–5 in. deep. High tensile steel rod was laid longitudinally down this channel and tied in with the rod starters and mesh of the hull. The wooden beams were then laminated to a constant camber on a prepared form and placed in position at 18 in. to 20 in. centres. Holes were drilled through the ends where they protruded into the ring beam cavity and a long steel holding rod was passed through. This rod was tied to the others in the beam shelf. After the laminates had been aligned and faired in with a long batten, a rich cement mortar was poured into the shuttering imbedding the ends of the beams completely. The plywood deck was fastened to the beams with glue and Gripfast nails, and trimmed off flush with the hull. Gaps beneath were filled with an epoxy resin cement paste. A wooden capping piece or rubbing stake was attached to the hull finishing off the deck edge. This was fastened in place with expanded bolts imbedded in holes in the hull. The deck was also sheathed in glassfibre. Although a somewhat spur-of-the-moment job, the structure has proved to be extremely strong and robust.

Fittings

A great deal has often been made in other publications of the difficulty of preparing holes in ferrocement to take skin fittings, with the stress being placed on the need to position all such apertures before the hull is plastered. This is completely untrue. Only where through hull openings are particularly wide, as in the case of portholes, windows, stern tube or rudder stock entry, is it advisable to make provision for them. Holes are easily made in the hull. Pilot holes can be drilled with a carborundum tipped masonry bit and a cold chisel and hammer will do the rest. Where rod reinforcement is encountered a hacksaw can be used to cut it, but attempt to position the hole between rods if possible. Skin fittings such as galley drain outlet, depth sounder, log, toilet inlet/outlets, etc can be fastened with screw flanges and glued in place with the tried and true epoxy resin.

Chainplates can be incorporated within the hull by welding them to the main reinforcement prior to plastering, or they can merely be tied in. Chainplates can be constructed from strips of $\frac{1}{4}$ in. steel, flared in a V at the bottom; sometimes a looped section of high tensile steel has been used, its legs imbedded in the hull shell. Imbedded plates could present problems later, so it might be a better idea to bolt the plates to the side of the hull. Bolt holes are easily drilled. Imbedded chainplates, if damaged by collision or corrosion, will be almost impossible to remove without damaging the hull.

Stanchion posts have been attached to a ferrocement deck with screws and rawl plugs, or even expanding bolts, but this does not provide a really strong fastening. Through-bolting is the best answer.

Considerable strain is going to be taken on the stemhead fitting, which may incorporate a chainplate and chain roller/fairlead. As a strong attachment is required it would be an idea to weld the fitting to the stem pipe or reinforcing rods at the bow. I know of one ferrocement yacht in which this fitting was merely through bolted to a thin ferrocement section. During a particularly heavy storm, while the boat was anchored in a bay on the East Coast, the rucking of the anchor chain pulled the fitting almost clear of the deck.

45 The development of a rudder for a yacht. Quarter inch steel plate was cut to shape, as seen. Wooden fairing strips were then bolted on horizontally, plywood cut to shape layered over them, screwed and glued. Glass cloth covers the construction to finish the job. The leading edge of the rudder was strengthened with further strips of $\frac{1}{4}$ in. plate.

Where port lights are being built into a hull they must be prepared before the hull is plastered. Openings must be cut to size and shape in the reinforcement. Try and site them in such a way to enable stringer rods and transversals to frame them. The actual frames of the ports can be fabricated out of steel plate and welded in place to the reinforcement. Before plastering a wooden blank can be fitted over them to keep them clear of mortar and allow the plasterer to trowel in to a neat edge. When these blanks are later removed a narrow flange is left to which the glass or plastic can be bolted. A rubber sealant or thin neoprene strip will make the joint watertight.

Rudder

Rudders can also be constructed in ferrocement. Like the rest of the hull they have to be lofted. The rudder stock can be made out of mild steel pipe, at either end of which a solid steel rod is inserted and welded in place. The solid section will in one instance fit into a bearing at the base of the sternpost; in the other it will form the rudder yoke which will pass into the hull to take a tiller or quadrant for steering. To this stock a framework of mild steel rod can now be welded to form the shape of the rudder. Over this two layers of mesh can be placed, wrapped round the leading edge of the rudder. Plastering can take place in the usual way with a vibrator being used to ensure proper penetration. When the mortar has cured further braces may be welded onto the stock and bolted to the ferrocement to stiffen and prevent any deflection of the rudder.

This form of construction is rather involved. A simpler technique was used in building a rudder for *Petralis*. In her case the base of the keel had been extended aft slightly during hull construction to provide a foot into which a bearing was fastened. This was kept clear during plastering. At the horn piece a pipe to take the rudder stock was permanently welded in place. The rudder was actually constructed in three sections. The main body consisted of a $\frac{1}{4}$ in. steel plate which had been cut to the necessary shape. The leading edge was stiffened with a 2 in. wide strip of $\frac{1}{4}$ in. plate which was welded in place. Instead of leaving this as a flat section the rudder was developed to provide a more efficient aerofoil shape. Wooden fairing strips were bolted to each side of the plate steel. Plywood was glued and nailed over these strips and eventually fibreglassed and painted with anti-fouling. The rudder stock was made out of two turned solid rods with flanges welded to them for the attachment of the rudder itself. Diameter of the stock was around 2 in. The bottom was fitted into the bearing attached to the heel of the vessel and the shaft at the top fitted into the pipe welded into the hull. The pipe rose well above the waterline. A further bearing was fitted to the top of this pipe to take the rudder stock. A keyway had been cut into the stock to take a quadrant or emergency tiller.

Repairs

Repairs are easily effected on a ferrocement hull. Damage tends to be localised to the point of impact, and can be repaired in different ways. Where a dent has been created this can often be bent back in place with a bit of force and the area faired in with the rest of the hull. Where cracking and chipping of the mortar has occurred, or a hole made, all the loose mortar must be knocked out. The area must be brushed clean of any dust and smaller particles, but large lumps of mortar which still cling to the mortar can be left. The reinforcement, if it has been bent out of alignment, must be hammered back in place. The hole can now be patched. The most effective means we have used is to coat the exposed surface with an epoxy resin and then plaster up the damage with an epoxy suitably stiffened with either dry cement or fine sand. Where damage has occurred below the waterline, there are some quick setting cement mortars on the market which could effectively be used.

BIBLIOGRAPHY

FERROCEMENT, WITH PARTICULAR REFERENCE TO MARINE APPLICATIONS, by Charles Darwin Canby. Paper read to NW section SNAME in March 1969

Ferrocement boat construction, by Jack L. Whitener, Cornell Maritime Press Inc., Cambridge, Maryland 1971

Concrete boatbuilding, its techniques and its future, by Gainor W. Jackson Jnr and Morley Sutherland, George Allen & Unwin Ltd, London 1969

How to build a ferrocement boat, by John Samson and Geoff Wellens, Samson Marine Design Enterprises Ltd, Ladner, BC, Canada 1968

YM AND FERRO, *Yachting Monthly* reprint 1969

Elements of yacht design, by Norman L. Skene, revised by Francis S. Kinney, Adam and Charles Black, London 1962

FERROCEMENT IN NEW ZEALAND, by Gainor W. Jackson Jnr, *Yachting Monthly* April 1971

REINFORCED CEMENT-MORTAR CONSTRUCTION, AN INVESTIGATION OF FERROCEMENT USING EXPANDED METAL, by J. G. Byrne and W. Wright. *Reinforced Cement-Mortar Construction Journal*, December 1961

FERROCEMENT SEA-GOING ARCHITECTURAL CONCRETE. *Concrete Trends,* Kaiser Cement, Oakland, California 1969

TESTING FERROCEMENT SAMPLES AT THE UNIVERSITY OF MICHIGAN. *National Fisherman,* October 1968

REPORT ON A NEW FERROCEMENT METHOD, by Samson Marine Design Enterprises, 1970

SOME EXPERIMENTS IN DESIGN AND CONSTRUCTION WITH FERROCEMENT by Lyal D. G. Collen, MAI, The Institution of Civil Engineers of Ireland, paper. January 1960

VAITSES LAUNCHES 64 FT CEMENT YACHT, by Burton T. Coffey. *National Fisherman*, April 1971

CIMENT FONDU FOR MAKING FERROCEMENT BOATS. Technical Service Section, Lafarge Aluminous Cement Co. Ltd. 1970

FERROCEMENT CONSTRUCTION, by T. L. James, Seacrete Ltd. Article in Southampton University Yacht Research Report No. 26, April 1969

TOP SEACRETE, Seacrete service and test report. Windboats Ltd, Wroxham, Norfolk. 1971

BLUE CIRCLE NEW PRODUCTS, leaflet, January 1970

CONCRETE AS A SHIPBUILDING MATERIAL, by R. G. Morgan, BSc, MICE. Paper read to International Marine and Shipping Conference, London, June 1969

FERRO TO LAST, by R. L. Waddell, P.Eng and T. W. Beckett. *Yachting Monthly,* 1969

LAMBOT'S BOATS, A PERSONAL REDISCOVERY, by W. Fisher Cassie CBE. Article in *Concrete,* Vol I No. 11, November 1967

FERROCEMENT: ITS CHARACTERISTICS AND POTENTIALITIES, by Prof. Pier Luigi Nervi. Translated from *L'Ingegnere* 1951, No. 1 by Cement and Concrete Association

TENTATIVE REQUIREMENTS FOR THE CONSTRUCTION OF YACHTS AND SMALL CRAFT IN FERROCEMENT. Lloyds Register of Shipping technical note, January 1967

COLVIN TRIES FERROCEMENT; DESCRIBES RESULTS. Article, *National Fisherman,* February 1968

THE MAKING OF A CONCRETE KETCH, by Dick Rath. Article, *Boating,* December 1968

FERROCEMENT BOATS. Concrete information from Cement and Concrete Association of Australia

FAO TO USE FERROCEMENT BOATS IN UAR FISHERY PROJECT'. Article in *Lloyd's Shipping List,* October 28, 1970

WIDE INTEREST SHOWN IN FERROCEMENT BOATS, by John Gardner. *National Fisherman,* September 1967

AUSSIES OUTLINE BOATBUILDING IN FERRO-CEMENT, by John Gardner. *National Fisherman,* October 1967

Boatbuilding with Hartley, by Richard Hartley, Boughtwood Printing House, Takapuna North, New Zealand 1967

FERROCEMENT FOR THE AMATEUR BUILDER, by Martin E. Iorns, consultant, Fibersteel Co., West Sacramento. Article, January 1971

CHINESE BUILD CONCRETE BOATS. *Concrete Products,* December 1966

FERROCEMENT BOATS. Concrete Report, Portland Cement Association, Illinois, 1969

FERRO PROGRESS, by Chris Cairncross. *Yachting Monthly,* March 1970

FERROCEMENT IN BOAT CONSTRUCTION. Article, *New Zealand Concrete Construction,* journal of NZ Portland Cement Association, May 1970

Reinforced concrete simply explained, by Oscar Faber. Oxford University Press.

Faber's reinforced concrete, by Faber and Mead. E. and F. N. Spon Ltd, 1961

Practical quality control for concrete, by Joseph J. Waddell. McGraw Hill

SHRINKAGE-CRACKING CHARACTERISTICS OF STRUCTURAL LIGHTWEIGHT CONCRETE, by R. G. McKeen and W. B. Ledbetter. *Journal of the American Concrete Institute,* October 1970

Concrete materials and practice, by L. J. Murdock and G. F. Blackledge. Edward Arnold

STRUCTURES, by P. L. Nervi, translated by F. W. Dodge Corp, New York.

Lightweight concrete, by Short and Kenniburgh. CR Books Ltd, London.

NEW SPLASH FOR CONCRETE. Article, *Concrete Construction*, November 1960

FEATHERSTONE AHOY. Article, *Concrete Construction*, July 1963

REINFORCED CONCRETE SHIPS, BARGES AND PONTOONS, by Alfred B. Searle. Article in *Concrete and Constructional Engineering* supplement *Concrete Shipbuilding*, December 1918

CEMENT BOATBUILDING PROBLEMS AIRED, by Martin E. Iorns, *National Fisherman*, May 1967

FAO TURNS TO FERROCEMENT BOATBUILDING. *Fishing News*, London, May 8, 1970

SEATTLE DESIGNER WRITES ON FERROCEMENT, by Arthur Drilling. *National Fisherman*, October 1967

FERROCEMENT MAKES STRONG HULL, by John Garner. *National Fisherman*, March 1967

MORE COUNTRIES TRY FERROCEMENT FOR FISHING BOATS. *Fishing News*, London, May 8, 1970

HOW SAND AND CEMENT WENT TO SEA AND MADE A CONCRETE BOAT, by Philip Clarke. *Sunday Times*, June 2, 1968

CONSTRUCTION OF A FERRO-SHOTCRETE MOTOR-SAILER HULL, by L. Lachance and P. Fugere. Article in *L'Ingenieur*, March 1970, Montreal

Practical ferro-cement boatbuilding, by Jay R. Benford and Herman Husen. International Marine Publishing Co., Camden, Maine, 1971

THE USE OF FERRO-CEMENT IN BOATBUILDING. Reprint on articles in *National Fisherman*, 1969